EAA Series

CW01455367

Editors-in-chief

Hansjoerg Albrecher	University of Lausanne, Lausanne, Switzerland
Ulrich Orbanz	University Salzburg, Salzburg, Austria

Editors

Michael Koller	ETH Zurich, Zurich, Switzerland
Ermanno Pitacco	Università di Trieste, Trieste, Italy
Christian Hipp	Universität Karlsruhe, Karlsruhe, Germany
Antoon Pelsser	Maastricht University, Maastricht, The Netherlands
Alexander J. McNeil	Heriot-Watt University, Edinburgh, UK

EAA series is successor of the EAA Lecture Notes and supported by the European Actuarial Academy (EAA GmbH), founded on the 29 August, 2005 in Cologne (Germany) by the Actuarial Associations of Austria, Germany, the Netherlands and Switzerland. EAA offers actuarial education including examination, permanent education for certified actuaries and consulting on actuarial education.

actuarial-academy.com

For further titles published in this series, please go to
http://www.springer.com/series/7879

Griselda Deelstra · Guillaume Plantin

Risk
Theory and
Reinsurance

Translated from the French by Urmie Ray

Springer

Griselda Deelstra
Département de Mathématique
Université Libre de Bruxelles
Brussels, Belgium

Guillaume Plantin
Manufacture des Tabacs
Toulouse School of Economics
Toulouse, France

Translated from the French language edition:
'Théorie du risque et réassurance'
by Griselda Deelstra and Guillaume Plantin
Copyright © Editions Economica 2005
Springer-Verlag London is part of Springer Science+Business Media
All Rights Reserved.

ISSN 1869-6929   ISSN 1869-6937 (electronic)
EAA Series
ISBN 978-1-4471-5567-6   ISBN 978-1-4471-5568-3 (eBook)
DOI 10.1007/978-1-4471-5568-3
Springer London Heidelberg New York Dordrecht

Library of Congress Control Number: 2013955490

Mathematics Subject Classification: 91B30, 62P05, 58E17

Springer is part of Springer Science+Business Media (www.springer.com)

Preface

Reinsurance is an important production factor of non-life insurance. The efficiency and the capacity of the reinsurance market directly regulate those of insurance markets. Indeed, the risks held by primary insurers are not maximally diversified. The latter tend to specialize geographically and technically to make their distribution structures profitable and exploit their underwriting expertise. Only reinsurance allows them to smooth out over space and time claim record spikes that would otherwise be too important for their capital. Reinsurers make a more intensive use of actuarial models than do primary insurers since they bear the most risky and the least diversifiable part of the claims. Actuarial models in reinsurance are based on a set of theoretical tools commonly referred to as "risk theory".

The purpose of this book is to provide a concise introduction to risk theory, as well as to its main application procedures to reinsurance.

The first part covers risk theory. After reminding the reader of insurance premium calculation principles and of mathematical tools enabling portfolios to be ordered according to their risk levels—"orders" on risks—it presents the most prevalent model of the process of total claim amounts generated by a portfolio, namely "the collective model". This first part ends with the main results of ruin theory. The purpose of ruin theory is to provide models for the occurrence of bankruptcy of an insurance company solely due to an extremely adverse claim record.

The second part describes the institutional context of reinsurance. It first strives to clarify the legal nature of reinsurance transactions. It next describes the structure of the reinsurance market, and then the different legal and technical features of reinsurance contracts, called reinsurance "treaties" by practitioners. Indeed, the business of reinsurance, though only lightly regulated, takes place within a set of customary rules, making it thereby easier to describe and understand. In particular, traditional reinsurance treaties fall into a limited number of broad categories.

The third part creates a link between the theories presented in the first part and the practice described in the second one. Indeed, it sets out, mostly through examples, some methods for pricing and optimizing reinsurance. Our aim is to apply the formalism presented in the first part to the institutional framework given in the second part. In particular, we try to highlight the fact that in many cases, the treaties used in practice correspond to the optimal risk sharing predicted by the theory. It is reassuring and interesting to find such a relation between approaches seemingly abstract and solutions adopted by practitioners.

As this book endeavours to provide a concise introduction to risk theory, as well as to its application procedures to reinsurance, it is mainly aimed at master's students in actuarial science, as well as at practitioners wishing to revive their knowledge of risk theory or to quickly learn about the main mechanisms of reinsurance. It is based on lecture notes, inspired by the Dutch text book of Kaas and Goovaerts (1993), for the course on "Risk Theory and Reinsurance", taught by both authors to third year students of the *Ecole Nationale de la Statistique et de l'Administration Economique* (E.N.S.A.E.). It is a translation of the French version entitled "Théorie du risque et réassurance", that has been published in 2006 by Economica in the Collection "Economie et statistiques avancées". Some references and data have been updated. We would like to express our sincere thanks to Christian Gouriéroux for his encouragements, his careful reading of that manuscript and for his detailed suggestions. We are also grateful to Rob Kaas for very helpful comments. All errors are ours.

Contents

Chapter 1
Elements of Risk Theory

1.1 Generalities on Risk Theory

Mathematical models in reinsurance and in non-life insurance are based on some fundamental tools. In this section, we give and describe their properties. Pure premium calculation principles and order relations on risks are successively discussed.

As these tools have already been set forth in a comprehensive and detailed manner in several books on actuarial science referenced in the bibliography, we intentionally only give a short overview. The aim is to allow the reader to find his way in this extensive literature, and to assemble the important results that will be referred to in the following sections. For more details on premium calculation principles, the reader can, for example, have a look at the books by Gerber (1979), Goovaerts et al. (1984) and Kaas et al. (2008), and for more details on orderings of risks, at the book by Goovaerts et al. (1990) or Kaas et al. (2008).

1.1.1 Premium Calculation Principles and Their Properties

Calculating an insurance or a reinsurance premium amounts to determining the exchange value of future and random cash flows. This means determining "the price of risk". Financial economy proposes two broad kinds of methods to carry out this evaluation: equilibrium valuation and arbitrage pricing. Equilibrium valuation is simply the application of the law of supply and demand to markets of future random flows. Arbitrage pricing derives the prices of some flows from the given prices of other flows by requiring that two portfolios that generate the same future flows have the same values. Unfortunately, for the time being, neither of these two approaches give satisfactory results in reinsurance.

Indeed, equilibrium valuations involve assumptions on buyers' and sellers' preferences which are impossible to formulate realistically for insurance or reinsurance companies as these are economic agents whose behaviour is complex. Within the framework of the best known and the simplest equilibrium valuation model, the

G. Deelstra, G. Plantin, *Risk Theory and Reinsurance*, EAA Series,
DOI 10.1007/978-1-4471-5568-3_1, © Springer-Verlag London 2014

Capital Asset Pricing Model described in most handbooks on finance (CAPM), (see, for example, Bodie and Merton 2001), such assumptions can be dispensed with if the risks are supposed to be Gaussian. This assumption is unfortunately not realistic in reinsurance.

Arbitrage pricing does not require such a complete model of the economy, but presupposes that all participants can buy or short sell risks at any time with reasonably low transaction costs, and that the information regarding risks is not too unequally distributed among agents. Unfortunately, these assumptions are very clearly not fulfilled in the reinsurance market.

Therefore, insurance and especially reinsurance premium calculation is not based on very deep theoretical foundations. A great diversity of methods exist alongside each other, without there being any truly decisive criteria to classify them. This section presents the ten most common ones, and then analyzes their main properties.

Major Premium Calculation Principles

Throughout this section, S will denote an almost-surely (a.s.) positive real random variable, modeling future insurance portfolio claim amounts, a portfolio possibly reduced to one contract, and F will be its (cumulative) distribution function (d.f.). $\mathfrak{F}$ will denote the set of distribution functions under consideration.

A premium calculation principle for a portfolio is a functional H

$$H : \mathfrak{F} \longrightarrow \mathbb{R}^+ \cup \{+\infty\}$$

$$F \mapsto H(F) = \Pi(S).$$

$\Pi(S)$ is therefore the amount of the premium associated to the risk S. A risk S is said to be insurable for the principle H if

$$H(F) = \Pi(S) < +\infty.$$

The ten most frequently mentioned premium calculation principles in actuarial literature are:

1. **The Zero Utility Principle**. This principle attributes a von Neumann Morgenstern utility function u to the reinsurer and computes the premium as the certainty equivalent of the risk underwritten, hence as the solution of the equation

$$E\left(u\left(\Pi_u(S) - S\right)\right) = u(0).$$

   Therefore, considering the reinsurer's risk aversion, $\Pi_u(S)$ is the minimum premium that he is ready to accept in order to cover the portfolio. Pratt's theorem says that if v is more concave than u ($v \circ u^{-1}$ is concave), then $\Pi_v \geq \Pi_u$.

Proof

$$E\left(v\left(\Pi_u(S) - S\right)\right) = E\left(v \circ u^{-1} \circ u\left(\Pi_u(S) - S\right)\right),$$

applying Jensen's inequality to $v \circ u^{-1}$:

$$E\left(v \circ u^{-1} \circ u\left(\Pi_u(S) - S\right)\right) \leq v \circ u^{-1}\left(E\left(u\left(\Pi_u(S) - S\right)\right)\right).$$

However,

$$v \circ u^{-1}\left(E\left(u\left(\Pi_u(S) - S\right)\right)\right) = v(0). \qquad \square$$

2. **The Pure Premium Principle.**

$$\Pi(S) = E(S).$$

The pure premium principle is obviously the special case of the zero utility principle where the reinsurer is supposed to be risk-neutral ($u(x) = x$).

3. **The Expected Value Principle.** Let $\beta > 0$,

$$\Pi(S) = (1 + \beta)E(S).$$

Other things being equal, the premium therefore increases with β. This principle adds a safety margin to the pure premium, which is merely proportional to the average risk level. It does not take into account other characteristics of its distribution.

4. **The Variance Principle.** Let $\beta > 0$,

$$\Pi(S) = E(S) + \beta \operatorname{Var}(S).$$

Here, the safety margin added to the pure premium is proportional to the variance of the risk. Note that if S is expressed in euros, then β is expressed in 1/euros.

5. **The Standard Deviation Principle.** Let $\beta > 0$,

$$\Pi(S) = E(S) + \beta\sqrt{\operatorname{Var}(S)}.$$

The safety margin also takes into account risk variability. It is, however, proportional to the standard deviation of the risk and not to its variance as in 4. Note that, as in 3, here β is dimensionless.

6. **The Exponential Principle.** Let $\alpha > 0$,

$$\Pi(S) = \frac{1}{\alpha} \ln\left(E\left(e^{\alpha S}\right)\right).$$

The parameter $\alpha > 0$ is called risk aversion. A premium calculated using the exponential principle increases with α and approaches the pure premium when $\alpha \to 0$. Indeed, the exponential principle is the principle of zero utility with an exponential utility function, for which α measures the reinsurer's absolute risk aversion index.

7. **The Swiss Principle**. Let f be a continuous, strictly increasing and convex function on $\mathbb{R}^+$ and $\alpha \in [0, 1]$. The premium is the solution of the equation

$$E\big(f\big(S - \alpha \Pi(S)\big)\big) = f\big((1 - \alpha)\Pi(S)\big).$$

If $\alpha = 1$, we again get the principle of zero utility by setting $u(x) = -f(-x)$.
8. **The Mean Value Principle**. This is the Swiss principle in the case $\alpha = 0$.
9. **The Maximal Loss Principle**. This is the "radical" principle that enables the reinsurer to secure a positive result!

$$\Pi(S) = \max(S).$$

It is a principle of zero utility with an infinitely risk averse reinsurer.
10. **The Esscher Principle**. Let $\alpha \geq 0$

$$\Pi(S) = \frac{E(Se^{\alpha S})}{E(e^{\alpha S})}.$$

The expected value S should, therefore, be recalculated under a new d.f. G, the "Esscher transform" of F, such that

$$dG(x) = \frac{e^{\alpha x}dF(x)}{\int e^{\alpha x}dF(x)}.$$

Thus, the most adverse states of nature are overweight relative to the objective probability.

Properties of Premium Calculation Principles

A natural criterion for deciding which premium calculation principle to adopt is to verify this principle by applying it to a number of properties of practical importance. The following four properties are important in practice. We state them, and then study the extent to which they are satisfied by the premium calculation principles defined above.

1. "**At least pure premium**". In order to ensure that the reinsurer makes a profit in expected value, the calculation principle must provide a result at least equal to that obtained by the pure premium principle.
2. **Translation invariance**. Adding a non-risky flow c to the portfolio implies that

$$\Pi(S + c) = \Pi(S) + c.$$

This property formalizes the intuitive fact that adding transfers of non-risky flows between insurer and insured should not modify the risk price.

Table 1.1 Properties of premium calculation principles

Principle	Prop. 1	Prop. 2	Prop. 3	Prop. 4
Zero utility	+	+	e	e
Pure premium	+	+	+	+
Expected value	+	−	+	−
Variance	+	+	+	−
Standard deviation	+	+	−	−
Exponential	+	+	+	+
Mean value	+	e	e	+
Maximum loss	+	+	+	+
Esscher	+	+	+	−

3. **Additivity**. For two *independent* portfolios S and S',

$$\Pi(S + S') = \Pi(S) + \Pi(S').$$

This property is questioned by a number of authors. For reasons of diversification, they consider that it would be preferable that the premium of the sum be lower than the sum of the premiums, as for example, when the standard deviation principle is applied.

4. **Iterativity**. For all S and S',

$$\Pi(S) = \Pi(\Pi(S \mid S')).$$

Suppose, for example, that the number of yearly accidents caused by a driver is distributed according to the Poisson law with parameter λ. The parameter λ is unknown and different for each driver. We suppose that λ is the realization of a real random variable (r.r.v.) Λ. Hence, the law of the number of accidents N conditioning on $\Lambda = \lambda$ is Poisson(λ). Iterativity says that the premium to be applied to the driver is obtained by applying the principle first to a claim record conditioning on Λ, and then to the random variable (function of Λ) thus obtained.

The consistency of each of the premium calculation principles with these four properties is summarized in Table 1.1. A "+" means that the property holds, a "−" that it does not. An "e" means it holds only in the exponential case.

The proof of the table is easy and left as an exercise using the various principles.

Except for the maximum loss principle, of no practical interest in general, the pure premium and the exponential principles

- are the only additive and iterative zero utility principles;
- are the only mean value principles invariant under translation and satisfying the additivity property.

This makes the exponential principle the only principle ensuring an expected non-zero profit and satisfying the four given properties.

Examples of Use

This section presents the first two simple examples using the exponential and the variance principles. The first example is taken from Gerber (1979) and the second one is based on Gouriéroux (1999).

(i) Optimal Coinsurance and the Exponential Principle

Coinsurance is an arrangement whereby several insurers share a risk, each one of them assuming a part of it for a part of the premiums. Consider n insurers, such that each insurer i sets prices according to the exponential principle with coefficient α_i, assuming a risk S in coinsurance. What is the optimal coinsurance? In other words, what part S_i should each of them take of S in order to minimize the total premium?
  Let α be such that

$$\frac{1}{\alpha} = \sum_{i=1}^{n} \frac{1}{\alpha_i}.$$

  Define the coinsurance $S_i^* = \frac{\alpha}{\alpha_i} S$ for all i. For this coinsurance, the premium is priced at

$$\Pi_{\min} = \frac{1}{\alpha} \ln\left(E e^{\alpha S}\right).$$

In fact, there is no coinsurance leading to a lower total premium: this coinsurance is optimal. Naturally, this coinsurance is such that each insurer receives a part which decreases with his risk aversion.

Proof According to Hölder's inequality, every coinsurance $(S_i)_{1 \le i \le N}$ satisfies

$$\prod_{i=1}^{n} E\left(\exp(\alpha_i S_i)\right)^{\frac{\alpha}{\alpha_i}} \ge E\left(\prod_{i=1}^{n} \exp(\alpha S_i)\right) = E e^{\alpha S}.$$

  Hence, $\frac{1}{\alpha} \ln E e^{\alpha S}$ is the minimum premium among the premiums calculated by using the exponential principle: $\sum_{i=1}^{n} \frac{1}{\alpha_i} \ln E e^{\alpha_i S_i}$, which implies optimality.   □

(ii) Reinsurance Cost and the Variance Principle

Suppose that an insurer insures a risk S, pricing it according to the variance principle. Suppose that a reinsurer reinsures this risk, pricing it according to the same principle. He proposes a contract in aggregate loss with priority y. This means that he pays the full amount of the total claim amounts exceeding the first y euros. The aim of this example is to show the impact of reinsurance cost on insurance cost.

Let Π be the premium that the primary insurer would apply if he was not reinsured. We get:

$$\Pi = ES + \beta \operatorname{Var} S.$$

Let Π_R be the premium actually set by the insurer when reinsured. Suppose that the full reinsurance cost is passed on to the client. Write

$$a(S) = \min(S, y) \quad \text{and} \quad r(S) = \max(S - y, 0)$$

for the insurer and the reinsurer's claim amounts to pay, respectively. Naturally,

$$a(S) + r(S) = S.$$

Then

$$\Pi_R = E(a) + \beta \operatorname{Var}(a)$$
$$+ E(r) + \beta \operatorname{Var}(r).$$

Equivalently,

$$\Pi_R = E(S) + \beta \operatorname{Var} S - 2\beta \operatorname{cov}(a, r).$$

However,

$$\operatorname{cov}(a, r) = E(ar) - E(a)E(r) = \big(y - E(a)\big)E(r) \geq 0$$

since

$$y \geq E(a) = E\big(\min(S, y)\big).$$

It follows that:

$$\Pi \geq \Pi_R.$$

1.1.2 Orderings of Risks

This section shows how the following question can be answered: "Is the risk S more risky than the risk S'?" Mathematically, it comes down to giving the principal (partial) order relations applied to sets of random variables. These are first- and second-order stochastic dominance relations. There are three equivalent definitions of a second-order stochastic dominance.

Throughout this section, we compare two positive real random variables S and S' modeling the claim amounts of a given portfolio over a given time period. Set F, F' and f, f' to be their respective d.f.'s and probability density functions.

(i) First-Order Stochastic Dominance

Definition 1 (First-Order Stochastic Dominance) S is preferred to S' in the sense of the first-order stochastic dominance (or simply "at first-order") iff for any increasing function u

$$Eu(-S) \geq Eu(-S').$$

In this case, we will write $S \geq_1 S'$.

In particular, the application of this definition to the identity function implies

$$S \geq_1 S' \quad \longrightarrow \quad ES \leq ES'.$$

If S dominates S', the expectation of the claim record associated to S is, therefore, necessarily less than that of S'.

The following proposition gives a characterization of first-order stochastic dominance by means of d.f.'s.

Proposition 2

$$S \geq_1 S' \quad \longleftrightarrow \quad F \geq F'.$$

Proof

$(\rightarrow)$ Set $u_y(x) = 1_{\{x \geq -y\}}$ for all y. u_y is an increasing function. Hence, $Eu_y(-S) \geq Eu_y(-S')$. However, $Eu_y(-S) = F(y)$, $Eu_y(-S') = F'(y)$.
$(\leftarrow)$ Use the fact that u is the monotone limit of increasing step functions. $\square$

This characterization provides clear insight into first-order stochastic dominance: a risk is preferred to another one if the probability that its magnitude be less than a given threshold y is greater, for any threshold y taken. Equivalently, it can be written $P(S \geq y) \leq P(S' \geq y)$ for all $y \geq 0$. This allows to interpret dominance at first order by means of the relative tail weights of distributions S and S'. In insurance, important risks are usually emphasized and the survival function is used: $\overline{F}_S(y) = P(S \geq y) = 1 - F_S(y)$.

The following proposition gives a sufficient condition for the first-order stochastic dominance involving the density functions of the claim size variables.

Proposition 3 (Single Crossing) *If there exists a positive real number c such that*

$$\begin{cases} f' \leq f & in \ ]0, c], \\ f' \geq f & in \ [c, +\infty[, \end{cases}$$

then $S \geq_1 S'$.

Proof In this case $F(0) - F'(0) \geq 0$ and $F - F'$ increases and then decreases to $\lim_{x \to \infty}(F(x) - F'(x)) = 0$, and $F - F'$ is, therefore, positive. $\square$

The following, so-called "coupling" proposition ensures that first-order stochastic dominance is preserved under convolution and composition of random variables.

Proposition 4 (Coupling)

(i) $S \le S'$ a.s. $\longrightarrow$ $S \ge_1 S'$.
(ii) *If $S \ge_1 S'$, then there exists a random variable S_1' with the same law as S' such that*

$$S \le S_1' \quad a.s.$$

(ii) is in some ways a "weak converse" of (i).

Proof

(i) is obvious (characterization by d.f.'s).
(ii) $S_1' = F'^{-1}(F(S))$ is a good candidate. Indeed

$$P\big(S_1' \le y\big) = P\big(S \le F^{-1}\big(F'(y)\big)\big) = F'(y)$$

and

$$F'^{-1}\big(F(x)\big) \ge x. \qquad \square$$

This coupling property enables us to easily obtain part of the following proposition, which will be extremely useful in the next sections.

Proposition 5 *Let $(S_i)_{i\in\mathbb{N}}$ and $(S_i')_{i\in\mathbb{N}}$ be two sequences of i.i.d. positive real random variables and N and N' two random variables with values in $\mathbb{N}$ such that $((S_i)_{i\in\mathbb{N}}, N)$, as well as $((S_i')_{i\in\mathbb{N}}, N')$ are independent.*

(i) *First-order stochastic dominance is preserved under convolution:*

$$S_1 \ge_1 S_2' \quad and \quad S_2 \ge_1 S_2' \quad \to \quad S_1 + S_2 \ge_1 S_1' + S_2'.$$

(ii) *First-order stochastic dominance is preserved under composition:*

$$\forall i \in \mathbb{N}, \quad S_i \ge_1 S_i' \quad and \quad N \ge_1 N' \quad \to \quad \sum_{i=1}^{N} S_i \ge_1 \sum_{i=1}^{N'} S_i'.$$

The result on convolutions follows immediately from the coupling property.

Let us now consider the monotonic property of various premium calculation principles when the space of risk variables is endowed with the ordering $\le_1$. We need to figure out whether the "less risky" risks at first-order lead to lower premiums. The following results hold:

1. The pure premium, the expected value and the zero utility principles (hence exponential) are such that

$$S \ge_1 S' \quad \longrightarrow \quad \Pi(S) \le \Pi\big(S'\big).$$

2. This property does not hold for the variance principle.

For the pure premium and expected value principles, the result follows from $ES \leq ES'$. For the zero utility principle, it is a direct consequence of the coupling property.

For the variance principle, the following is a counterexample. Suppose that S takes value 0 with probability $1 - p$ and value 1 with probability p, whereas S' is deterministic and takes value 1. $S \geq_1 S'$, but

$$\Pi(S) = p + \beta p(1 - p) > \Pi(S'),$$

whenever $\beta p > 1$.

(ii) Second-Order Stochastic Dominance or Stop-Loss Orders

Let us first define the following three partial orders:

Definition 6 (*RA* Order, or Ordering Induced by All Risk Averse Individuals) S is preferred to S' in the sense of the ordering induced by all risk averse individuals iff

$$Eu(-S) \geq Eu(-S')$$

for any increasing concave function u.

Definition 7 (*SL* Order, or Stop-Loss Order) S is preferred to S' with respect to the stop-loss order iff

$$\forall d > 0, \quad \Pi_S(d) \leq \Pi_{S'}(d)$$

where $\Pi_X(d) = E((X - d)^+) = \int_d^{+\infty}(x - d)dG(x)$ is the "stop-loss transform" of the variable X with d.f. G.

d can be interpreted as the deductible level the risk-taker leaves the insured to pay on each claim. In this case $\Pi_S(d) = E((S - d)^+)$ is the claim cost for the risk-taker. The risk S is then preferable to the risk S' with respect to the stop-loss order if the average cost for S is lower for all possible deductible levels.

Definition 8 (*V* Order, or Variability Order) S is preferable to S' with respect to the variability order iff there exists a random variable S'' such that

(i) $S + S''$ and S' are equidistributed.
(ii) $E(S'' \mid S) \geq 0$ a.s.

This definition implies that a risk S is preferred to another risk S' if this risk S' can be interpreted as the sum of S and of the result of a random experiment with a positive conditional expected value.

The following is a relatively strong result.

Proposition 9 *RA, SL and V are identical.*

Proof We will only prove $RA \rightarrow SL$, $SL \rightarrow RA$ and $V \rightarrow SL$. The proof of $SL \rightarrow V$ is longer and beyond the scope of this section, whose purpose is to give an overview. It can, for example, be found in Rothschild and Stiglitz (1970).

$$RA \rightarrow SL:$$

For all $d > 0$, $x \mapsto \max(x - d, 0)$ is increasing and convex.

$$SL \rightarrow RA:$$

Any increasing convex function is the monotone limit of combinations of linear functions of this type.

$$V \rightarrow SL:$$

For all $d > 0$,

$$
\begin{aligned}
E\big((S' - d)^+\big) &= E\big((S + S'' - d)^+\big) \\
&= E\big(E\big((S + S'' - d)^+ \mid S\big)\big) \\
&\geq E\big((S + E(S'' \mid S) - d)^+\big) \\
&\geq E\big((S - d)^+\big).
\end{aligned}
$$ □

As a result of this equivalence, the terms "stop-loss order", dear to actuaries, or "second-order stochastic dominance", dear to economists, will be interchangeably used to qualify this order on risks, and will be denoted $\leq_2$.

This order is widely used in economy (information theory) and in finance (pricing criteria in incomplete markets).

The definition RA gives good insight into this ordering. S is preferred to S' if all risk averse people prefer S to S'. It also shows that first-order stochastic dominance implies second-order stochastic dominance.

The following proposition contains a sufficient condition, practical for second-order stochastic dominance using d.f.'s.

Proposition 10 (Single Crossing) *If*

(i) $ES \leq ES'$,
(ii) *There exists $c \geq 0$ s.t.*

$$
\begin{cases}
F \leq F' & in\ [0, c], \\
F \geq F' & in\ [c, +\infty],
\end{cases}
$$

then $S \geq_2 S'$.

Proof Applying definition SL, the result follows immediately since

$$\Pi_S(d) = \int_d^{+\infty} \big(1 - F(y)\big)dy.$$ □

The following proposition will be widely used in the next sections. We assume it holds. The proof can be found in Goovaerts et al. (1990).

Proposition 11 *Let* $(S_i)_{i \in \mathbb{N}}$ *and* $(S'_i)_{i \in \mathbb{N}}$ *be two sequences of i.i.d. positive real random variables and* N *and* N' *two random variables with values in* $\mathbb{N}$ *such that* $((S_i)_{i \in \mathbb{N}}, N)$ *and* $((S'_i)_{i \in \mathbb{N}}, N')$ *are independent.*

(i) *Second-order stochastic dominance is preserved under convolution*:

$$S_1 \geq_2 S'_2 \quad and \quad S_2 \geq_2 S'_2 \quad \rightarrow \quad S_1 + S_2 \geq_2 S'_1 + S'_2.$$

(ii) *Second-order stochastic dominance is preserved under composition*:

$$\forall i \in \mathbb{N}, \quad S_i \geq_2 S'_i \quad and \quad N \geq_2 N' \quad \rightarrow \quad \sum_{i=1}^{N} S_i \geq_2 \sum_{i=1}^{N'} S'_i.$$

To illustrate this result, we prove the following corollary:

$$X \geq_2 Y \quad \rightarrow \quad X + Z \geq_2 Y + Z$$

for Z independent of X and Y.

The proof uses definition SL:

$$E(X + Z - d)^+ = E\big(E(X + Z - d \mid Z)^+\big)$$

$$= \int E\big(X - (d - z)\big)^+ dF_Z(z)$$

$$\leq \int E\big(Y - (d - z)\big)^+ dF_Z(z)$$

$$= E(Y + Z - d)^+.$$

We conclude this section with a specific example of a family of risk variables having a maximal and a minimal element with respect to the stop-loss order.

This is the set of real variables having a common support $[0, b]$ as well as a common expected value μ. The minimal element of this family with respect to the stop-loss order is the deterministic variable equal to μ. The maximal element is the binomial variable Y such that

$$P(Y = b) = \frac{\mu}{b} = 1 - P(Y = 0).$$

To convince himself of this, the reader only needs to draw the graph of the d.f.'s of these two variables and notice that all the other d.f.'s of this family only crosses them once.

This little example illustrates the idea, according to which, the heavier the tails of its distribution, the more risky a variable at second order.

1.2 The Collective Model

The term "collective model" stands for all models of the aggregate claims of a port-folio within a given time interval as the composition of a random variable repre-senting the individual claim amount and of a counting variable, called a "frequency variable" (a random variable with values in $\mathbb{N}$ counting the number of claims). This model is at the heart of the entire mathematics of non-life insurance. In practice, among these models, those whose frequency laws are Poisson laws prevail.

That this model has come to prevail is naturally in large part due both to a certain amount of realism and to satisfactory results in terms of estimates. But, as illustrated by the following example, workability arguments also strongly argue for this choice.

1.2.1 The Individual Model Versus the Collective Model

Suppose that a portfolio consists of n independent risks. Each risk $i \in \{1, \ldots, n\}$ gives rise to a claim amount, for instance annual, X_i, whose d.f. has been estimated to be equal to F_i.

A reasonable model for the annual total claim amount S of the portfolio is

$$S = \sum_{i=1}^{n} X_i.$$

This is the individual model.

In particular, if risks are equidistributed and n is large, the central limit theo-rem, recalled below in its simplest form, allows us to assert that a normal law is a reasonable model for the law of S.

Proposition 12 (Central Limit Theorem) *Let* $(X_i)_{i \in \{1, \ldots, n\}}$ *be n independent and identically distributed real random variables with finite expected value and vari-ance, written* μ *and* σ^2 *respectively.* $\sum_{i=1}^{n} \frac{(X_i - \mu)}{\sigma \sqrt{n}}$ *converges in law to a standard normal variable.*

Extensions of the central limit theorem allow to partially weaken the assumptions of independence and equidistribution.

If n is not too large, it is easy to estimate the distribution of S as the convolution of the n distributions of the variables X_i. For large n, relatively heavy and sophisticated simulation methods are necessary.

Let us now consider the following alternative model $\widetilde{S} = \sum_{i=1}^{n} \widetilde{X}_i$ with $\forall i \in \{1, \ldots, n\}$

$$\widetilde{X}_i = \sum_{j=1}^{v_i} Z_{ij}$$

where v_i follows a Poisson law with parameter 1 and for all i, the Z_{ij} are independent and equidistributed according to F_i. The variables (Z_{ij}, v_i) are assumed to be independent.

Three arguments show its superiority to the individual model, regardless of any quality criteria for the estimate.

1. Since second-order stochastic dominance is preserved under composition and convolution, it is dominated by the individual model at second order. It is therefore more "prudent" in the sense that it gives a more risky representation of the portfolio.
2. Compounded Poisson laws are preserved under convolution. More precisely, $\widetilde{S}$ follows a compound Poisson law whose counting variable is a Poisson law with parameter n and the claim size law has d.f. $\frac{1}{n}\sum_i F_i$. Indeed, the generating function of $\widetilde{S}$ equals:

$$E e^{t\widetilde{S}} = E e^{t(\sum_{i=1}^{n} \widetilde{X}_i)} = \prod_{i=1}^{n} E e^{t\widetilde{X}_i}$$

$$= \prod_{i=1}^{n} E e^{t(\sum_{j=1}^{v_i} Z_{ij})} = \prod_{i=1}^{n} E\left[E\left(e^{t(\sum_{j=1}^{v_i} Z_{ij})} \mid v_i\right)\right]$$

$$= \prod_{i=1}^{n} E\left(E\left(e^{tX_i}\right)^{v_i}\right) = \prod_{i=1}^{n} e^{E(e^{tX_i}-1)}$$

$$= e^{n(\frac{1}{n}\sum_{i=1}^{n} E e^{tX_i}-1)}.$$

3. There are methods for approximating the d.f. of a compound Poisson variable that can be efficiently implemented.

Thus, the collective model, compared to the individual model, provides a much more manageable mathematical representation of the total portfolio claim amount, but not at the cost of prudence.

However simplistic, in our opinion, this short example illustrates the fundamental reasons why the collective model, in particular when Poisson compound, actually stands out in non-life insurance. The proof of the model's importance is the "frequency/average cost" reasoning adopted by all insurance practitioners, actuaries and others, to analyze the total claim amount of a portfolio. It should, nonetheless, be noted that as computing powers increase, simulation methods are becoming more and more attractive and common, while they are less restrictive from the modeling side.

This section recalls the main results regarding the collective model, both static and dynamic. The first two moments as well as the moment generating function of the total claim amount are computed. The main specifications adopted in practice for the estimate of the frequency and of the cost are discussed. Finally, in the case of compound Poisson models, two methods are given for computing the d.f. of the total claim amount.

1.2.2 Dynamic Collective Model Versus Static Collective Model

For the modeling of the total claim amount related to a portfolio, two strategies can be adopted.

A rather modest strategy is a *static* modeling. A given time t_0 is fixed and in this case, only the *variable S* representing the aggregate claims at t_0 is modeled, without bothering about what happens after t_0. In this case, the collective model is

$$S = \sum_{i=1}^{N} X_i,$$

where the individual claim sizes (X_i) are independent and equidistributed, and N is an independent counting variable, also called a frequency variable, in other words a random variable with values in $\mathbb{N}$ representing the number of claims arising.

A more ambitious strategy is a *dynamic* modeling of the stochastic *process* $(S_t)_{t \geq 0}$ of the total claim amount. The dynamic model in the manner of Lundberg considers $(S_t)_{t \geq 0}$ as a family of real random variables indexed by $t \in [0, \infty[$ such that each S_t represents the accumulated claims during the period 0 to t. In this case, it is assumed that the total claim amount S_t can be written

$$S_t = \sum_{i=1}^{N_t} X_i$$

where the individual claim sizes (X_i) in the time period 0 to t are independent and equidistributed, and N_t is an independent counting process, in other words a family of random variables with values in $\mathbb{N}$ indexed by $t \in [0, \infty[$, such that N_t is the number of claims arising in the time period 0 to t.

In what follows, we will mostly study the dynamic collective model in the manner of Lundberg. It is the most general model. It can be used to study the long term evolution of an insurance company. Moreover, naturally, if such a dynamic model is adopted, then we also have a static modeling of the accumulated claims in any time period $[t_1, t_2]$. The counting variable is then distributed as $N_{t_2} - N_{t_1}$ conditioned upon N_{t_1}.

1.2.3 Typical Properties of the Collective Model

Let

$$S = \sum_{i=1}^{N} X_i$$

denote a static collective model. The individual claim size variables $(X_i)_{i \in \mathbb{N}}$ are i.i.d. and the variables $\{(X_i)_{i \in \mathbb{N}}, N\}$ are independent. The first two moments of S can be simply expressed.

Proposition 13

$$ES = EN \times EX.$$

Proof

$$ES = E\left(E\left(\sum_{i=1}^{N} X_i \mid N\right)\right) = E\big(N \times (EX)\big) = EN \times EX.$$

□

Proposition 14

$$\text{Var } S = EN \times \text{Var } X + (EX)^2 \times \text{Var } N.$$

Proof

$$E\big(S^2 \mid N\big) = E\left(\sum_{i=1}^{N} X_i^2 + 2 \sum_{1 \le i < j \le N} X_i X_j\right)$$

$$= NE\big(X^2\big) + N(N-1)(EX)^2.$$

Hence, $E(S^2) = E(N^2)(EX)^2 + EN(E(X^2) - (EX)^2)$, and

$$\text{Var } S = EN \times \text{Var } X + (EX)^2 \times \text{Var } N.$$

□

Now, recall that the moment generating function M_X of a random variable X is defined by

$$M_X(t) = E\big(e^{tX}\big)$$

where this expected value exists. In the case of the collective model, it follows that

Proposition 15

$$M_S = M_N \circ \ln(M_X).$$

Proof

$$Ee^{tS} = E\big(E\big(e^{tS} \mid N\big)\big)$$

$$= E\big(\big(Ee^{tX}\big)^N\big) = E\big(e^{N \ln(Ee^{tX})}\big).$$

□

Note that Propositions 13 and 14 follow from the previous one: compute the first two derivatives of the moment generating function at 0.

Finally recall that the d.f. of S can be written as

$$F_S(x) = \sum_{n \in \mathbb{N}} F^{*n}(x) \times P(N = n),$$

where $*n$ denotes the n-fold convolution of a d.f. This formula has little practical value since, as stated in the introduction of this section, a great advantage of the collective model is that it avoids heavy computations of convolutions.

The following two sections list the main specifications used to calibrate the collective model. Statistical methods that can be practically implemented to choose or estimate a specification are beyond the subject of this book. The interested reader may refer to any book on punctual and collective estimation.

1.2.4 Principal Frequency Models Used in Practice

Three types of frequency laws are used in practice, depending on the respective position of the variance and of the expected value of the frequency.

(i) The Variance Is Significantly Smaller than the Expected Value: Binomial Law

Recall that if the frequency of the claims incurred by a portfolio of n risks follows a binomial law with parameter p, then the probability that the number of claims equals $k \in \{0, \ldots, n\}$ is

$$\frac{n!}{(n-k)!k!}(1-p)^{n-k}p^k.$$

(ii) The Variance Is Equivalent to the Expected Value: Poisson Process or Poisson Law

Poisson laws and Poisson processes play a central role in non-life actuarial science, in particular because of the simple formulas they give rise to. Here we directly present the dynamic modeling using Poisson processes. Static models where the frequency law is a Poisson law can be directly deduced.

Poisson Processes

Let us recall the definition and the main properties of Poisson processes. We adopt the presentation given by Lamberton and Lapeyre (1991) which is an excellent trade-off between concision and completeness.

Definition 16 (Poisson Process) Let $(T_i)_{i \in \mathbb{N}}$ be a sequence of i.i.d. real random variables having an exponential law with parameter λ, and density:

$$1_{\{x>0\}} \lambda e^{-\lambda x}.$$

Let

$$\tau_n = \sum_{i=1}^{n} T_i.$$

A Poisson process of intensity λ is the process N_t defined by

$$N_t = \sum_{n \geq 1} 1_{\{\tau_n \leq t\}} = \sum_{n \geq 1} n 1_{\{\tau_n \leq t < \tau_{n+1}\}}.$$

Hence, N_t is the number of points of the sequence $(\tau_n)_{n \geq 1}$ less than or equal to t.

Proposition 17 (The Poisson Process Follows a Poisson Law) *If $(N_t)_{t \geq 0}$ is a Poisson process of intensity λ, then for all $t > 0$ N_t follows a Poisson law with parameter λ:*

$$P(N_t = n) = e^{-\lambda t} \frac{(\lambda t)^n}{n!}, \quad n \in \mathbb{N}$$

and

$$E N_t = \operatorname{Var} N_t = \lambda t,$$
$$E\left(e^{s N_t}\right) = e^{\lambda t (e^s - 1)}.$$

Proposition 18 *Let $(N_t)_{t \geq 0}$ be a Poisson process of intensity λ. $(N_t)_{t \geq 0}$ is a process with stationary and independent increments.*

Proposition 19 (Two Characterizations of a Poisson Process)

(i) *$(N_t)_{t \geq 0}$ is a Poisson process of intensity λ iff it is a homogeneous right-continuous and left-limited Markov process such that*

$$P(N_t = n) = e^{-\lambda t} \frac{(\lambda t)^n}{n!}.$$

(ii) *$(N_t)_{t \geq 0}$ is a Poisson process iff it is a right-continuous increasing process with stationary and independent increments, increasing only by jumps of 1.*

Modeling the counting process by a Poisson process is of interest because of the remarkable properties of the compound Poisson processes, given here.

Compound Poisson Processes

A compound Poisson process is a dynamic collective model whose counting process is a Poisson process.

A first property of practical interest concerning compound Poisson processes is that the variance of the total claim amount at time t takes an extremely simple form:

$$\operatorname{Var}(S_t) = \lambda t \times E\left(X^2\right).$$

A second deeper property is stability under convolution of the family of compound Poisson processes.

Proposition 20 (Stability Under Convolution of Compound Poisson Processes) *If $(S_i)_{1 \leq i \leq m}$ are m independent compound Poisson processes with Poisson parameters $(\lambda_i)_{1 \leq i \leq m}$ respectively, and whose individual claim sizes follow the d.f.'s $(F_i)_{1 \leq i \leq m}$, respectively, then*

$$S = \sum_{i=1}^{m} S_i$$

is a compound Poisson process with Poisson parameter

$$\lambda = \sum_{i=1}^{m} \lambda_i$$

and d.f. F s.t.

$$F(x) = \sum_{i=1}^{m} \frac{\lambda_i}{\lambda} F_i(x).$$

Proof The proof follows easily from the relation between the characteristic functions $\varphi_S(t) = E[e^{itS}]$ and $\varphi_j(t) = E[e^{itX_j}]$:

$$\varphi_S(t) = \prod_{i=1}^{m} e^{\lambda_i(\varphi_i(t)-1)} = e^{\lambda(\sum_{i=1}^{m} \frac{\lambda_i}{\lambda} \varphi_i(t)-1)}.$$

$\square$

Finally, the following proposition, given without proof (see Feller 1969), shows that by changing the time scale, each counting process with stationary and independent increments can be transformed into a Poisson process.

Proposition 21 (Operational Time) *Let $(N_t)_{t \geq 0}$ be a counting process with stationary and independent increments. If $EN_t = m(t)$ is an increasing continuous function, then the process defined by*

$$\overline{N_\tau} = N_{t(\tau)}$$

with

$$t(\tau) = \inf\{s \mid m(s) = \tau\}$$

is a Poisson process with parameter 1.

$m(t)$ is called the operational time. It can prove useful for if, say over a year, the number of claims is not distributed according to a Poisson law, a time change can bring us back to a Poisson process.

(iii) The Variance Is Significantly Higher than the Expected Value: Mixed Poisson Law, in Particular Negative Binomial Law

A mixed Poisson law can be considered as a Poisson law whose intensity λ is random, and having a d.f. U called a "risk structure function". For such a law, we therefore have:

$$P(N_t = n) = \int e^{-\theta t} \frac{(\theta t)^n}{n!} dU(\theta),$$

$$EN = t E\lambda,$$

$$\text{Var } N_t = t E\lambda + t^2 \text{ Var } \lambda.$$

In insurance, the mixed Poisson law mostly used in practice is the negative binomial law, for which the risk structure function is a Gamma law with density

$$U'(\lambda) = \frac{c^\gamma}{\Gamma(\gamma)} \lambda^{\gamma-1} e^{-c\lambda},$$

where $c > 0$, $\gamma > 1$ and $\Gamma(\gamma) = \int_0^{+\infty} u^{\gamma-1} e^{-u} du$.

In the case of the negative binomial law,

$$P(N_t = k) = \frac{\Gamma(\gamma + k)}{\Gamma(\gamma)k!} p^\gamma (1 - p)^k, \quad k \in \mathbb{N}$$

where

$$p = \frac{c}{c + t},$$

and

$$E N_t = \frac{\gamma t}{c},$$

$$\text{Var } N_t = \frac{\gamma t}{c} + \gamma \left(\frac{t}{c}\right)^2.$$

Therefore, compared to the Poisson law, there is an extra t^2 term added to the variance.

1.2.5 Principal Claim Size Distributions Used in Practice

We will only list them without explaining in detail the empirical and theoretical reasons that can lead to choose one over the other. This question could in itself fill an entire book (see, for example, Hogg and Klugman 1984) and does not fall within the scope of this book, kept deliberately concise. The definitions of the nine laws that we give are based on their density f or on their d.f. F.

1. Exponential Law

$$\lambda > 0, \quad f(x) = 1_{\{x>0\}} \lambda e^{-\lambda x}.$$

2. Gamma Law

$$\alpha, \beta > 0, \quad f(x) = 1_{\{x>0\}} \frac{\beta^{\alpha}}{\Gamma(\alpha)} x^{\alpha-1} e^{-\beta x}.$$

3. Weibull Law

$$c > 0, \tau \geq 1, \quad f(x) = 1_{\{x>0\}} c \tau x^{\tau-1} e^{-cx^{\tau}}.$$

4. Lognormal Law

$$\sigma > 0, \quad f(x) = 1_{\{x>0\}} \frac{1}{\sqrt{2\pi}\sigma x} e^{-\frac{(\ln x - m)^2}{2\sigma^2}}.$$

5. Pareto Law

$$s, \alpha > 0, \quad f(x) = 1_{\{x>s\}} \frac{a}{x} \left(\frac{s}{x}\right)^a.$$

6. Burr Law

$$\alpha, \kappa, \tau > 0, \quad 1 - F(x) = 1_{\{x>0\}} \left(\frac{\kappa}{\kappa + x^{\tau}}\right)^{\alpha}.$$

7. Benktander Law of the First Kind

$$\alpha, \beta > 0, \quad 1 - F(x) = 1_{\{x\geq1\}} \left(1 + 2\frac{\beta}{\alpha} \ln x\right) e^{-\beta(\ln x)^2 - (1+\alpha)\ln x}.$$

8. Benktander Law of the Second Kind

$$\alpha > 0, 0 < \beta \leq 1, \quad 1 - F(x) = 1_{\{x>1\}} e^{\frac{\alpha}{\beta}} x^{\beta-1} e^{-\alpha \frac{x^{\beta}}{\beta}}.$$

9. Log-Gamma Law

$$\alpha, \beta > 0, \quad 1 - F(x) = 1_{\{x\geq1\}} \frac{\alpha^{\beta}}{\Gamma(\beta)} (\ln x)^{\beta-1} x^{-\alpha-1}.$$

As an example, let us compute the average claim size for an exponential and lognormal distribution, when the contract includes a deductible equal to K.

(1) In the case of an exponential distribution with parameter λ, the stop-loss premium is easily computed:

$$E\left[(X - K)^+\right] = \frac{1}{\lambda} e^{-\lambda K}.$$

It can be seen that if the retention K increases, then the premium decreases exponentially. If λ decreases and hence the mean of X increases (since

$E[X] = \frac{1}{\lambda}$), the premium increases. An exponential specification is frequently used to model an unknown *duration*, such as the time it takes for a borrower to default in credit insurance.

(2) If X follows a lognormal law such that $\ln X \frown N(m, \sigma^2)$, then a simple calculation shows that

$$E\left[(X - K)^+\right] = e^{m+\sigma^2/2} \Phi(d_1) - K\Phi(d_2) \qquad (*)$$

where $d_1 = \frac{m - \ln K + \sigma^2}{\sigma}$, $d_2 = d_1 - \sigma$, and Φ is the distribution function of a standard normal law.

It is easy to see that $\frac{\partial}{\partial m} E[(X - K)^+] = e^{m+\sigma^2/2}\Phi(d_1) \geq 0$ and to conclude that the stop-loss premium increases with the mean m.

We leave the reader to check as an exercise that the derivative of the stop-loss with respect to the standard deviation σ also shows that the stop-loss premium increases with the variance σ^2.

Since Merton and Black & Scholes' classic work (see Merton 1990 and Black and Scholes 1973), it is quite usual to model the asset price S_t at time t using a lognormal law under a risk-neutral probability

$$\ln S_t \frown N\left(\ln S_0 + \left(r - \frac{\sigma^2}{2}\right)t, \frac{\sigma^2}{2}t\right)$$

with r the short-term riskless interest rate. Formula (*) leads immediately to the price of a European call option C_t with strike price K. A European call option with strike price K gives the right to buy the underlying (a certain asset with price process $(S_t)_t$) at the fixed strike price K at a due date T. Under the lognormal assumption, the price of a European call option is given by the famous and frequently used Black-Scholes formula:

$$C_t = e^{-r(T-t)} E_Q\left[(S_T - K)^+ | \mathcal{F}_t\right] = S_t \Phi\left(d_1(t)\right) - K e^{-r(T-t)} \Phi\left(d_2(t)\right)$$

where

$$d_1(t) = \frac{\ln(S_t/K) + (r + \sigma^2/2)(T - t)}{\sigma\sqrt{T - t}},$$

$$d_2(t) = d_1(t) - \sigma\sqrt{T - t}.$$

Thanks to this explicit formula, it is easy to study price sensitivity to the parameters. The different derivatives of the option price are in practice very important in financial management. Their practical properties and usefulness are discussed in a separate chapter in any primer on derivative products (see e.g. Hull 2000).

The deviation towards finance in this example shows that the mathematical formalism of options in finance and of contracts with deductibles and ceilings in reinsurance is very similar. This similarity will reveal very useful in the last part of the present book.

1.2.6 *Two Methods to Compute the D.F. of a Collective Model*

As stated in the Introduction, among several models, the collective model has come to prevail in non-life insurance (and hence in reinsurance) for important workability reasons. Here we give two methods for the computation of the d.f. of a collective model, far easier than the computation, even approximate, of the convolution of a large number of random variables.

(i) The Panjer Recursive Algorithm

Thanks to this algorithm, the d.f. of the total claim amount can be efficiently computed when the individual claim size is modeled by a discrete random variable. The d.f. can therefore be approximated in the general case by discretizing the d.f. of the claim size. We state here only the algorithm since a clear proof can be found in many textbooks on risk theory, see e.g. Panjer and Willmot (1992).

Proposition 22 (The Panjer Recursive Algorithm) *Let $S = \sum_{i=1}^{N} X_i$ be a compound variable, such that the claim size variable is discrete and set*

$$P(X = x) = p(x).$$

Suppose there exist two reals a and b such that

$$\forall n \in \mathbb{N}\backslash\{0\}, \quad P(N = n) = q_n = \left(a + \frac{b}{n}\right)q_{n-1}.$$

Then $f(s) = P(S = s)$ satisfies

$$f(0) = \begin{cases} P(N = 0), & \text{if } p(0) = 0, \\ M_N(\ln p(0)), & \text{if } p(0) > 0, \end{cases}$$

$$\forall s \in \mathbb{N}\backslash\{0\}, \quad f(s) = \frac{1}{1 - ap(0)} \sum_{h=1}^{s} \left(a + \frac{bh}{s}\right) p(h) f(s - h).$$

The frequency laws satisfying the recursive property needed to implement the Panjer algorithm are:

- Poisson Law. For a Poisson law of intensity λ, $a = 0$ and $b = \lambda$. The Panjer formula becomes

$$f(0) = e^{-\lambda(1 - p(0))},$$

$$f(s) = \frac{1}{s} \sum_{h=1}^{s} \lambda h p(h) f(s - h).$$

- Negative Binomial Law.
- Binomial Law. For a binomial law (n, p), $p = \frac{a}{a-1}$ and $n = -\frac{b+a}{a}$.

Application   As an application of the Panjer algorithm, we will recursively calculate the stop-loss transform $E[(S-d)^+]$ of a discrete variable for any value of d. This application can be found in Kaas et al. (2008).

This stop-loss transform can be computed by one of the equivalent expressions:

$$E\big[(S-d)^+\big] = \int_d^\infty (x-d)\,dF_S(x)$$

$$= E[S] - d + \int_0^d (d-x)\,dF_S(x)$$

$$= \int_d^\infty \big(1 - F_S(x)\big)\,dx$$

$$= E[S] - \int_0^d \big(1 - F_S(x)\big)\,dx.$$

If S is a r.v. with integral values, the integrals can be replaced by series:

$$E\big[(S-d)^+\big] = \sum_{x=d}^\infty (x-d)P[S=x]$$

$$= E[S] - d + \sum_{x=0}^{d-1}(d-x)P[S=x]$$

$$= \sum_{x=d}^\infty \big(1 - F_S(x)\big)$$

$$= E[S] - \sum_{x=0}^{d-1}\big(1 - F_S(x)\big).$$

The last equality implies immediately a recursive relation between $\Pi(d) = E[(S-d)^+]$ and $\Pi(d-1)$:

$$\Pi(d) = \Pi(d-1) - \big(1 - F_S(d-1)\big)$$

where $\Pi(0) = E[S]$.

Then let S be the r.v. having a compound Poisson law with

$$P[X=1] = p(1) = \frac{1}{2} = p(2) = P(X=2) \quad \text{and} \quad \lambda = 1.$$

In this case, the Panjer algorithm simplifies to:

$$f(x) = \frac{1}{x}\left[\frac{1}{2}f(x-1) + f(x-2)\right], \quad x = 1, 2, \ldots$$

with $f(0) = e^{-1} = 0.368$, $F(0) = f(0)$, $\Pi(0) = E[S] = \lambda\mu_1 = 1.5$.

Table 1.2 Application: stop-loss transform

x	$f(x)$	$F(x) = F(x-1) + f(x)$	$\pi(x) = \pi(x-1) - 1 + F(x-1)$
0	0.368	0.368	1.500
1	0.184	0.552	0.868
2	0.230	0.782	0.420
3	0.100	0.881	0.201
4	0.070	0.951	0.083
5	0.027	0.978	0.034

The density, the d.f. and the stop-loss transform can then be recursively computed.

Calculations for the stop-loss transform for different retentions are summarized in Table 1.2.

(ii) Edgeworth Expansion

For simplicity's sake, let $S(t) = \sum_{j=1}^{N_t} Y_j$ be a compound Poisson process of fixed intensity 1.

The Edgeworth expansion method gives an asymptotic expansion of the d.f. of

$$X_N(\tau) = \frac{\tau EY - S(\tau)}{\sqrt{\tau E(Y^2)}},$$

$X_N(\tau)$ being the reduced centered total claim amount at time τ.

For a random variable Z, let us recall its characteristic function $\varphi_Z(t) = Ee^{itZ}$. Then

$$\varphi_{X_N(\tau)}(t) = Ee^{itX_N(\tau)}$$

$$= \exp\left(\tau \left(\frac{iE(Y)}{\sqrt{\tau E(Y^2)}} t - 1 + \varphi_Y\left(-\frac{t}{\sqrt{\tau E(Y^2)}} \right) \right) \right).$$

Expand $\ln(\varphi_{X_N(\tau)}(t))$ with respect to $\frac{1}{\tau}$, for example up to the order $\frac{3}{2}$

$$\ln\left(\varphi_{X_N(\tau)}(t)\right)$$

$$= \tau \left(\sum_{n=0}^{4} \frac{i^n (-1)^n t^n}{n!} \times \frac{E(Y^n)}{\sqrt{\tau E(Y^2)}^n} + \frac{iE(Y)}{\sqrt{\tau E(Y^2)}} t - 1 \right)$$

$$= -\frac{1}{2} t^2 + \frac{c_3}{3!\sqrt{\tau}} (-it)^3 + \frac{c_4}{4!\tau} (-it)^4 + 0\left(\tau^{-\frac{3}{2}}\right),$$

with

$$c_n = \frac{E(Y^n)}{\sqrt{E(Y^2)^n}}.$$

Take the exponential and expand the exponential up to the order $\frac{3}{2}$ of $(\frac{1}{\tau})$

$$\varphi_{X_N(\tau)}(t) = \int e^{itx} dF_{X_N(\tau)}(x)$$

$$= e^{-\frac{1}{2}t^2}\left(1 + \frac{c_3}{3!\sqrt{\tau}}(-it)^3 + \frac{c_4}{4!\tau}(-it)^4 + \frac{10c_3^2}{6!\tau}(-it)^6 + 0(\tau^{-\frac{3}{2}})\right).$$

Let us now recall that, if the nth derivative of the density of a reduced centered Gaussian variable is written $N^{(n)}(.)$,

$$\int e^{itx} dN^{(n)}(x) = (-it)^n e^{-\frac{1}{2}t^2}.$$

Identifying the integrands in the previous equation leads to

$$F_{X_N(\tau)}(x) = N(x) + \frac{c_3}{3!\sqrt{\tau}} N^{(3)}(x) + \frac{c_4}{4!\tau} N^{(4)}(x)$$

$$+ \frac{10c_3^2}{6!\tau} N^{(6)}(x) + 0(\tau^{-\frac{3}{2}}).$$

This corresponds to the desired result. Naturally, the same method can be applied to get higher order expansions.

The book by e.g. Beard et al. (1984) gives other methods for asymptotic expansions such as the "Normal Power" one or the Esscher expansion.

1.3 Ruin Theory

1.3.1 Introduction: Reinsurance and Solvency of Insurance Companies

Like any company, an insurance company is insolvent when it owes the rest of the world more than it is owed by the rest of the world. The size of the bankruptcy is then the difference between the two amounts. Under given conditions upon the initial capital and the operating cash flows of a company, the time and size of the next bankruptcy are naturally two random variables. Ruin theory aims to study the connections between these initial conditions and the distributions of these variables.

Before going into the details of ruin theory, this introduction briefly recalls why solvency in insurance and banking receives more attention than in other industries, and why in the context of insurance, reinsurance and solvency are two closely connected questions.

Why does solvency of insurance companies receive special attention in every country?

Financial institutions, banks or insurance companies are virtually the only companies whose solvency is subject to comprehensive regulations and special supervision by public authorities. In the case of a bank, a justification frequently given for this regulation is the contagion and panic effect in the event of its bankruptcy. Financial history is littered with "bank runs" by depositors of all banks in an economy in case of difficulties of a single isolated bank, whose consequences on the payment system and then on the real economy could be significant. However, examples of contagions of such magnitude are nonexistent in insurance.

The special social role of insurance is sometimes given to justify the regulation of solvency. This argument too is not clear: in light of the social consequences of the Enron bankruptcy, can those of the bankruptcy of the smallest mutual insurance company in Texas really be considered as too much to bear?

The most convincing justification for the prudential supervision of insurance companies is the *representation hypothesis* given by Dewatripont and Tirole (1994). It is based on the similarity between the mechanism of loan agreements and that of prudential regulations. On the one hand, loan agreements given by banks to their industrial customers contain provisions close to those of prudential regulations (maximum debt ratio, minimum liquidity, restrictions on the use of funds). On the other, banks are entitled to request the early repayment of their contribution when these rules are contravened. The difference between an industrial and an insurance company is that the latter's debt is held by scattered economic agents who are not especially knowledgeable about financial matters, namely the insured. The regulatory authority is therefore responsible for representing them and for taking, in their stead, the decision of "early repayment" or of liquidation, which is optimal when the solvency ratio is crossed, similarly to a bank deciding to liquidate the assets of a borrower in a financially delicate position.

Why is reinsurance an important determining factor in the solvency of an insurance company?

Reinsurance will be strictly defined in the second part of the book. At this stage, it suffices to say that reinsurance allows insurance companies to insure themselves, to their turn, against the risks they insure. To illustrate the connection between reinsurance and solvency, let us consider the following elementary model. An insurance company disposing of K amount of capital, underwrites N independent and identically distributed risks. Each risk gives rise to a random claim amount with finite expected value and standard deviation, written E and σ, respectively. The risks are priced according to the expected value principle, each premium being worth $(1 + \rho)E$. Set S to be the aggregate claims of the company.

The probability of ruin PR of the company equals

$$PR = P\big(K + N(1 + \rho)E - S < 0\big)$$
$$= P(S - NE > K + N\rho E)$$

From the Bienaymé-Tchebychev inequality, according to which for all $\lambda > 0$: $P(|S - E[S]| > \lambda) \leq \frac{\text{Var}[S]}{\lambda^2}$, it follows that

$$PR \leq \frac{1}{\beta^2}$$

where

$$\beta = \frac{K + N\rho E}{\sqrt{N} \times \sigma}$$

is *the safety coefficient*.

The Bienaymé-Tchebychev inequality gives a very large upper bound for the probability of ruin. In this example intended to illustrate, this is sufficient, but far more precise upper bounds can be found if the law of claim sizes is more specified.

The higher the safety coefficient of the company, the lower its probability of ruin. To increase the safety coefficient, under a given risk structure, it is possible

1. to increase its capital K;
2. to increase the number of contracts N (for $N > \frac{K}{\rho E}$, β increases with N);
3. to increase the price, in other words ρ.

For a given capital, the last two solutions are tricky to implement. Increasing the price degrades competitiveness and must therefore ultimately reduce the number of contracts, leading to a decrease of the safety coefficient. Increasing rapidly the number of contracts, even by means other than pricing, is a perilous exercise in insurance. Indeed, the most flexible part of the demand consists in general of the risks that are the hardest to estimate. The structure of risk can therefore be adversely altered.

Therefore, for a given capital, the only way to adjust the safety coefficient in the short term is to act, through reinsurance, directly on the risk structure without altering the portfolio. It is worth noting that if reinsurance implies a transfer of risk (diminution of σ), it also implies a transfer of profits (diminution of ρ). Determining an optimal reinsurance strategy is to arbitrate between these two effects, one positive, the other negative. The last chapter, which deals with reinsurance optimization, provides such arbitrations.

1.3.2 The Cramer-Lundberg Model

The model from which the main results of ruin theory are derived is the Cramer-Lundberg model (which goes back to Lundberg 1903). This model is set out in most (introductory) books on risk theory, for example in Bowers et al. (1986), Bühlmann (1970), Gerber (1979), Grandell (1991), Straub (1988), Beard et al. (1984), Daykin et al. (1994), Goovaerts et al. (1990), Heilmann (1988), Prabhu (1980), Kaas and Goovaerts (1993), Kaas et al. (2008), Mikosh (2009), Rolski et al. (1998), Tse (2009).

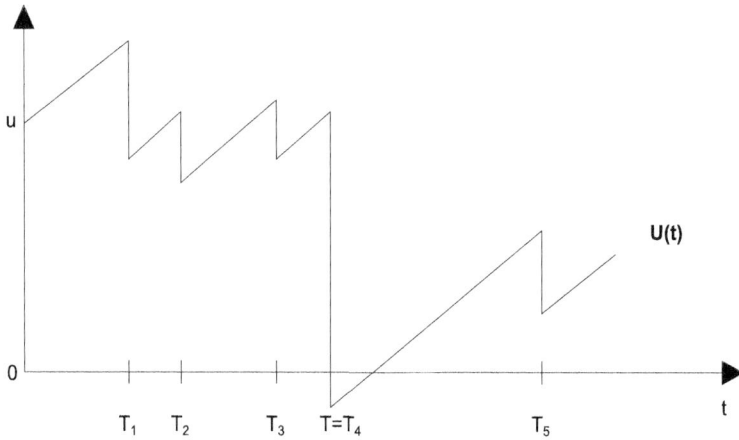

Fig. 1.1 A trajectory including ruin

The Cramer-Lundberg model represents the net value U_t of an insurance company at time t by the following risk process, usually called the free reserve or surplus process:

$$U_t = u + ct - S_t$$

where

- u is the initial capital at $t = 0$.
- $S_t = \sum_{i=1}^{N_t} X_i$ is the process of the total claim amount paid until time t. It is a compound Poisson process. The intensity of the Poisson process is written λ. The individual claim size admits an expected value μ and a d.f. F. We also write $M_X(s) = E(e^{sX})$.
- $c = (1 + \theta)\lambda\mu$ is the instantaneous rate of premiums received by the company. θ is the loading rate of pure premiums.

By definition, the company is ruined when

$$U_t < 0.$$

The aim of this chapter is to study the distribution of the real random variable

$$T = \inf\{t \mid U_t < 0\},$$

which is called the (possibly infinite) time of ruin. Figure 1.1 presents an example of a trajectory including ruin.

Write

$$\psi(u, t) = P(T < t)$$

for the probability of being ruined before time t, when the initial surplus is u.

$$\psi(u) = P(T < \infty)$$

is the limit of this probability (if it exists) when t tends to infinity.

Fig. 1.2 The Lundberg
coefficient R

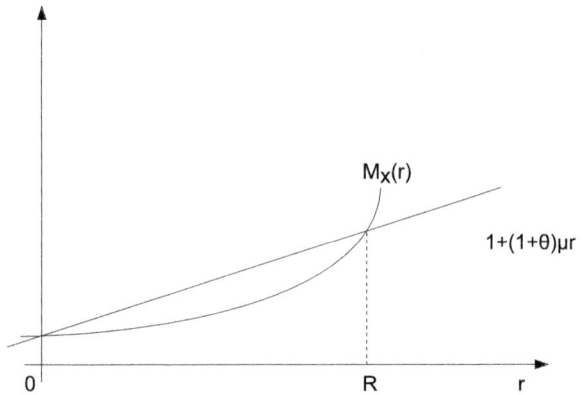

An important number is the adjustment or Lundberg coefficient R.

Definition 23 (Adjustment Coefficient or Lundberg Coefficient) The Lundberg coefficient is the strictly positive solution, if it exists, of the equation

$$1 + (1 + \theta)\mu r = M_X(r).$$

See Fig. 1.2 for a visualisation of this definition. The equation defining the Lundberg coefficient implicitly takes the following equivalent forms:

$$\lambda + cr = \lambda M_X(r),$$

$$\int \left(e^{rx} - 1 - \theta \right)\left(1 - F_X(x) \right)dx = 0,$$

$$e^{rc} = E e^{rS},$$

$$c = \frac{1}{r}\ln M_S(r).$$

Note that if the insurance company obtains a premium rate c by setting the price
according to the exponential principle with a coefficient α, then

$$R = \alpha.$$

1.3.3 The Lundberg Inequality

The Lundberg coefficient, when it exists, allows us to obtain an upper bound for the
probability of ruin.

Proposition 24 (Lundberg Inequality)

$$\psi(u, t) \le \psi(u) \le e^{-Ru}.$$

Proof We give a proof developed in Gerber (1979). It is very concise and elegant, and is based on the notion of a martingale. In informal terms, a stochastic process is a martingale if its value at a given time is the best prediction that can be made of its value at any future time.

Let $(\mathcal{F}_t)_{t\geq 0}$ be the filtration generated by $(U_t)_{t\geq 0}$. Let us first show that e^{-RU_t} is a martingale:

$$E\left(e^{-RU_{t+s}} \mid \mathcal{F}_t\right) = E\left(e^{-RU_t} \times e^{-R(cs-(S_{t+s}-S_t))} \mid \mathcal{F}_t\right)$$

$$= e^{-RU_t} \times e^{-Rcs} \times Ee^{RS_s}$$

$$= e^{-RU_t}.$$

This last equality follows from the definition of R. Next note that the time of ruin $T = \inf\{t \mid U_t < 0\}$ is a stopping time for $(\mathcal{F}_t)_{t\geq 0}$.

Hence, according to the stopping time theorem, $e^{-RU_{T\wedge t}}$ is also a martingale. Moreover,

$$1_{\{T\leq t\}} \leq e^{-RU_{T\wedge t}}.$$

Taking the expected value of this inequality gives as expected

$$\psi(u,t) = E 1_{\{T\leq t\}} \leq Ee^{-RU_{T\wedge t}} = e^{-Ru}. \qquad \square$$

The following proposition provides an equality for the probability of ruin based on the Lundberg coefficient.

Proposition 25 (An Equality for the Probability of Ruin)

$$\psi(u) = \frac{e^{-Ru}}{E\left(e^{-RU_T} \mid T < \infty\right)}.$$

Proof

$$\forall t \geq 0,$$

$$\underbrace{Ee^{-RU_t}}_{\substack{=e^{-Ru} \\ \text{(martingale)}}} = \underbrace{E\left(e^{-RU_t} \mid T \leq t\right)P(T \leq t)}_{=E(e^{-RU_T}\mid T\leq t)\psi(u,t)} + \underbrace{E\left(e^{-RU_t} \mid T > t\right)P(T > t)}_{\substack{\to 0 \\ t\to\infty}}.$$

We get the stated equality by passing to the limit for t tending to infinity. Let us show that the right-hand most term effectively approaches 0 when t tends to infinity. Let

$$u_0(t) = \beta t^{\frac{2}{3}}$$

with β a constant, chosen small enough. Then

$$E\left(e^{-RU_t} \mid T > t\right) P(T > t)$$

$$= E\left(e^{-RU_t} \mid T > t \text{ and } 0 \le U_t \le u_0(t)\right) P\left(T > t \text{ and } 0 \le U_t \le u_0(t)\right)$$

$$+ E\left(e^{-RU_t} \mid T > t \text{ and } U_t > u_0(t)\right) P\left(T > t \text{ and } U_t > u_0(t)\right)$$

$$\le P\left(U_t \le u_0(t)\right) + \underbrace{E\left(e^{-Ru_0(t)}\right)}_{\substack{\to 0 \\ t \to \infty}},$$

and

$$P\left(U_t \le u_0(t)\right) \underset{t \to \infty}{\to} 0.$$

Indeed, the expected value and the variance of U_t are linear in t. The Bienaymé-Tchebychev inequality therefore gives an upper bound for this probability, which approaches 0 as t approaches infinity.   □

Remarks

1. In particular, this equality gives the Lundberg inequality.
2. If the individual claim size, X, follows an exponential law with parameter $\frac{1}{\mu}$, then the above equality takes a particularly simple form:

$$\psi(u) = \frac{1}{1+\theta} e^{-\frac{\theta}{1+\theta} \times \frac{u}{\mu}}.$$

1.3.4 The Beekman Convolution Formula

The Beekman convolution formula for the probability of ruin is of interest because it does not depend on the existence of the Lundberg coefficient. By somewhat anticipating on the following section, this is very useful when the tail of the distribution of the individual claim size is fat. In this case, the Lundberg coefficient actually does not exist.

The strategy here is not to directly look for information on the distribution of T, but on the maximum aggregate loss (see e.g. Beekman 1974):

$$L = \max_t(S_t - ct).$$

This variable is relevant since ruin happens whenever the maximum aggregate loss becomes greater than the initial surplus u. The probability of ruin is thus given by

$$\psi(u) = 1 - F_L(u),$$

where F_L is the d.f. of the maximum aggregate loss. The probability that the company goes bankrupt is equal to the probability that the maximum aggregate loss is greater than the initial capital u.

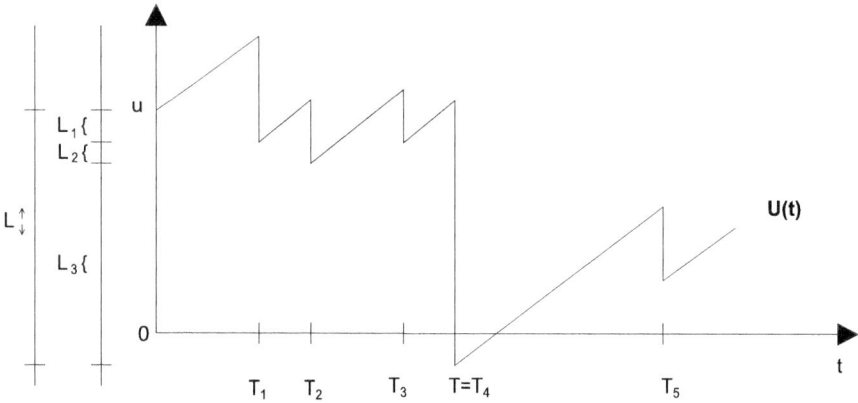

Fig. 1.3 The maximum aggregate loss

We have (see Fig. 1.3 for a visualisation)

$$L = \sum_{i=1}^{M} L_i$$

where

- L_i is the difference between the ith and $i + 1$th lowest historical levels reached by the surplus process before ruin, and is called "record low".
- M is the number of lowest historical levels reached before ruin.

Since a compound Poisson process has stationary and independent increments, the probability of reaching a new record low is the same each time a new record low is reached, and equals $\psi(0)$. Hence, M follows a geometric law with parameter $(1 - \psi(0))$:

$$P(M = m) = \psi(0)^m \left(1 - \psi(0)\right).$$

Therefore L is a compound geometric process. To obtain the law of L, the laws of each L_i remain to be characterized. It is given as a consequence of the following proposition:

Proposition 26 *For all $y > 0$:*

$$P\left(U(T) \in]-\infty, -y] \text{ and } T < \infty \mid U(0) = 0\right) = \frac{\lambda}{c} \int_y^{\infty} \left(1 - F(u)\right) du.$$

Proof We borrow this proof from Kaas et al. (2008). Let

$$G(u, y) = P\left(U_T \in]-\infty, -y] \text{ and } T < \infty \mid U(0) = u\right).$$

$G(u, y)$ is the probability that ruin happens and the deficit is worse than y when the initial capital equals u.

An infinitesimal reasoning in the time period t to $t + dt$ allows us to write

$$G(u, y)$$
$$= \underbrace{(1 - \lambda dt)G(u + cdt, y)}_{(1)}$$
$$+ \lambda dt \Bigg[\underbrace{\int_0^{u+cdt} G(u + cdt - x, y)dF(x) + \int_{u+y+cdt}^{\infty} dF(x)}_{(2)} \underbrace{\vphantom{\int}}_{(3)} \Bigg].$$

(1) corresponds to the situation where no claims occur in the time period t to $t + dt$.
(2) corresponds to the situation where a claim occurs in the time period t to $t + dt$, but its amount is too low to lead to ruin.
(3) corresponds to the situation where ruin occurs in the time period t to $t + dt$.

Divide the two terms of this equality by cdt and let dt approach 0. This gives:

$$\frac{\partial G}{\partial u}(u, y) = \frac{\lambda}{c}\left(G(u, y) - \int_0^u G(u - x, y)dF(x) - \int_{u+y}^{\infty} dF(x)\right).$$

Integrate with respect to u in the interval 0 to z

$$G(z, y) - G(0, y)$$
$$= \frac{\lambda}{c}\int_0^z G(u, y)du - \frac{\lambda}{c}\int_0^z \int_0^u G(u - x, y)dF(x)du$$
$$- \frac{\lambda}{c}\int_0^z \int_{u+y}^{\infty} dF(x)du.$$

However,

$$\int_0^z \int_0^u G(u - x, y)dF(x)du = \int_0^z \int_0^{z-v} G(v, y)dF(x)dv$$
$$= \int_0^z G(v, y)F(z - v)dv,$$

and

$$\int_0^z \int_{u+y}^{\infty} dF(x)du = \int_y^{z+y} \left(1 - F(v)\right)dv.$$

Hence,

$$G(z, y) - G(0, y)$$
$$= \frac{\lambda}{c}\left(\int_0^z G(u, y)\left(1 - F(z - u)\right)du - \int_y^{z+y} \left(1 - F(v)\right)dv\right).$$

The proposition follows by taking the limit of z to $+\infty$ in this equality.   □

Applying the previous proposition, we get

$$\psi(0) = G(0,0) = \frac{1}{1+\theta},$$

$$1 - F_{L_1}(y) = \frac{1}{\mu} \int_y^\infty \big(1 - F(x)\big)dx$$

since

$$1 - F_{L_1}(y) = \frac{G(0,y)}{\psi(0)}.$$

This allows us to state the Beekman convolution formula:

Proposition 27 (Beekman Convolution Formula)

$$\psi(u) = \sum_{m=0}^\infty p(1-p)^m \big(1 - F_I(u)^{*m}\big)$$

where

- *$*m$ is the m-fold convolution*
- $p = \frac{\theta}{1+\theta}$
- $F_I(x) = \frac{1}{\mu} \int_0^x (1 - F(y))dy.$

Proof It suffices to note that L has a compound geometric law with parameter p. □

Another application is the following formula for the moment generating function of the variable L:

$$M_L(r) = \frac{\theta}{1+\theta} + \frac{1}{1+\theta} \times \frac{\theta(M_X(r) - 1)}{1 + (1+\theta)\mu r - M_X(r)}.$$

Indeed, since $L = \sum_{i=1}^M L_i$ and M follows a geometric law with parameter $p = \frac{\theta}{1+\theta}$, the moment generating function for L is

$$M_L(r) = M_M\big(\ln(M_{L_1}(r))\big) = \frac{p}{1 - (1-p)M_{L_1}(r)},$$

and knowing that

$$M_{L_1}(r) = \frac{1}{\mu} \int e^{ry}\big(1 - F_X(y)\big)dy,$$

the conclusion follows.

Based on a functional equation, the limit of the probability of ruin when the initial capital become arbitrarily large follows from the last stated result (cf. Embrechts et al. 1997). Such a limit case is in practice interesting for several companies whose statistical ruin, under given conditions of operating, is highly unlikely in view of their capital.

1.3.5 Application of the Smith Renewal Theorem to Ruin Theory

To find the limit of the probability of ruin when the initial surplus becomes arbitrarily large, we apply the following theorem, which we state without proof.

Proposition 28 (Smith Renewal Theorem) *Let g be the solution of the functional equation*

$$g(t) = h(t) + \int_0^t g(t - x) dF(x) \quad for\ t \geq 0$$

where $\int_0^{+\infty} x dF(x) < \infty$ and h is the difference of increasing Riemann-integrable functions.
    Then

$$\lim_{t \to \infty} g(t) = \frac{\int_0^\infty h(x) dx}{\int_0^\infty x dF(x)}.$$

The purpose of this section is to show that the probability of ruin is the solution of such a functional equation in order to deduce the following result:

Proposition 29 (A Limit for the Probability of Ruin) *If the Lundberg coefficient R exists and if*

$$\int_0^\infty x e^{Rx} \big(1 - F(x)\big) dx < \infty,$$

then

$$\lim_{u \to \infty} e^{Ru} \psi(u) = \frac{\theta \mu}{R \int_0^\infty x e^{Rx} (1 - F(x)) dx} < \infty.$$

Proof

1. Let us first show the following equality

$$\psi(x) = \int_0^\infty \lambda e^{-\lambda t} \int_0^\infty \psi(x + ct - y) dF(y) dt.$$

Indeed, if $\psi_k(u)$ is the probability that ruin occurs at the latest when the kth claim occurs, then

$$\psi_k(u) = \int_0^\infty \int_0^\infty \psi_{k-1}(u + ct - x) dF(x) \lambda e^{-\lambda t} dt.$$

Letting k approach $+\infty$ gives the desired equality.
2. Deduce that

$$\psi(t) = \frac{\lambda}{c} \int_0^t \psi(t - y)\big(1 - F(y)\big) dy + \frac{\lambda}{c} \int_t^\infty \big(1 - F(y)\big) dy.$$

Changing to the variable $s = x + ct$ in the equality proved in 1 gives:

$$\psi(x) = \frac{\lambda}{c} \int_x^\infty \lambda e^{-\lambda \frac{(s-x)}{c}} \int_0^\infty \psi(s-y) dF(y) ds.$$

Take the derivative with respect to x:

$$\psi'(x) = \frac{\lambda}{c} \psi(x) - \frac{\lambda}{c} \int_0^\infty \psi(x-y) dF(y).$$

Equivalently

$$\psi'(x) = \frac{\lambda}{c} \psi(x) - \frac{\lambda}{c} \int_0^x \psi(x-y) dF(y) - \frac{\lambda}{c} (1 - F(x)).$$

Integrate this equality between 0 and t:

$$\psi(t) - \psi(0) = \frac{\lambda}{c} \int_0^t \psi(x) dx - \frac{\lambda}{c} \int_0^t \int_0^x \psi(x-y) dF(y) dx$$

$$- \frac{\lambda}{c} \int_0^t (1 - F(x)) dx,$$

and

$$\int_0^t \int_0^x \psi(x-y) dF(y) dx$$

$$= \int_0^t \int_y^t \psi(x-y) dx dF(y) \quad \text{(by reversing the order of integration)}$$

$$= \int_0^t \int_0^{t-y} \psi(u) du \, dF(y) \quad \text{(by a change of variable)}$$

$$= \int_0^t \psi(u) \int_0^{t-u} dF(y) du \quad \text{(by reversing the order of integration)}$$

$$= \int_0^t \psi(t-y) F(y) dy \quad \text{(by a change of variable)}.$$

This therefore gives the desired equality since $\psi(0) = \frac{\theta}{1+\theta}$.

3. Finally note that:

$$F_I(x) = \frac{1}{\mu} \int_0^x (1 - F(u)) du.$$

The preceding equality can be written

$$e^{Ru} \psi(u) = \frac{1}{1+\theta} e^{Ru} (1 - F_I(u)) + \int_0^u e^{R(u-x)} \psi(u-x) dF_{I,R}(x)$$

where

$$dF_{I,R}(x) = e^{Rx} d\left(\frac{1}{1+\theta} F_I(x)\right)$$

is an Esscher transform of F_I.

The result is then obtained by applying the Smith renewal theorem and by noting that

$$\int_0^\infty \frac{1}{1+\theta} e^{Ru}\left(1 - F_I(u)\right) du = \frac{\theta}{(1+\theta)R},$$

$$\int_0^\infty x\, dF_{I,R}(x) = \frac{1}{(1+\theta)\mu} \int_0^\infty x e^{Rx}\left(1 - F(x)\right) dx. \qquad \square$$

This result leads to the following equivalent expression for the probability of ruin when the initial capital is very large:

$$\psi(u) \sim_\infty C e^{-Ru},$$

with

$$C = \frac{\theta\mu}{R \int_0^\infty x e^{Rx}(1 - F(x))dx}.$$

1.3.6 Probability of Ruin and Orderings of Risks

Let us conclude this chapter with two results showing that the probability of ruin is a "monotone" function when the space of individual claim sizes is equipped with the stop-loss order (see Goovaerts et al. 1990 or Kaas et al. 2008).

We consider two Cramer-Lundberg models that are independent and identical in all respects except for the individual claim size variables, written X for one of the models and Y for the other.

Proposition 30 (Stop-Loss Order and Lundberg Coefficients) *If the two Lundberg coefficients R_X and R_Y are well defined, then*

$$X \geq_2 Y \quad \rightarrow \quad R_X \geq R_Y.$$

Proof The moment generating function of X is less than that of Y. $\qquad \square$

Proposition 31 (Stop-Loss Order and Probability of Ruin)

$$X \geq_2 Y \quad \rightarrow \quad \forall u \geq 0, \qquad \psi_X(u) \leq \psi_Y(u).$$

Proof Let us start by proving the following result:

If $X \geq_2 Y$, then $\exists D_1, D_2$ such that Y and $X + D_1 + D_2$ have the same distribution with $E(D_1 \mid X) = 0$ and $D_2 \geq 0$ almost surely.

For this, introduce

$$Y' = IY$$

where I is a Bernoulli variable such that

$$P(I = 1 \mid X = x, Y = y) = \frac{x}{E(Y \mid X = x)} = 1 - P(I = 0 \mid X = x, Y = y).$$

Then

$$Y' \leq Y$$

and

$$E(Y' \mid X) = X$$

by construction. Hence,

$$D_1 = Y' - X$$

$$D_2 = Y - Y'$$

satisfy the desired property. Set $Y'' = X + D_1 + D_2$.

First, it is clear that

$$\psi_{Y'}(u) \leq \psi_{Y''}(u)$$

since $D_2 \geq 0$ almost surely.

Moreover,

$$X \geq_2 Y' \quad \text{and} \quad EX = EY'$$

However,

$$1 - F^X_{L_i}(y) = \frac{\Pi_X(y)}{EX},$$

$$1 - F^{Y'}_{L_i}(y) = \frac{\Pi_{Y'}(y)}{EY'},$$

where $F^X_{L_i}$ and $F^{Y'}_{L_i}$ are the d.f.'s of the record lows defined in the section on the Beekman convolution for individual claim sizes equal to X and Y', respectively.

Hence,

$$L^X_i \geq_1 L^{Y'}_i.$$

The first-order stochastic dominance being preserved under composition, this inequality holds for maximum aggregate losses. This implies

$$\psi_X(u) \leq \psi_{Y'}(u)$$

and hence the desired result.   □

1.4 Risk Theory of Fat-Tailed Laws

When the Lundberg Coefficient Does not Exist   In the Cramer-Lundberg model, the equation implicitly defining the Lundberg coefficient has a non-negative solution only if

$$\int_0^\infty e^{rx} dF(x)$$

is defined at least in the neighborhood of 0. A necessary condition for the existence of this integral is

$$1 - F(x) \le e^{-rx} E\left(e^{rX}\right).$$

This means that the tail of the distribution of the individual claim size, X, must have an exponential bound, hence that important claim sizes must be sufficiently "rare". Unfortunately, this property is not satisfied by laws better adapted to risks sold on the reinsurance market.

For example, in the case of a Pareto law with parameter $\alpha > 1$:

$$\int_0^\infty e^{rx} (1+x)^{-\alpha} dx$$

does not converge for any $r > 0$.

This section gives, mostly without proof (see Embrechts et al. (1997) for a detailed presentation), some methods to deal with "fat-tailed" laws, thanks to which the non existence of the Lundberg coefficient can be surmounted. We only allude briefly to a literature in full expansion.

It is worth recalling that thanks to the Beekman convolution formula given in the preceding section, it is possible to understand the probability of ruin without taking account of the existence of the adjustment coefficient. Therefore, here, the strategy is to first "classify" the fat-tailed laws according to the fatness of their tails, and then to derive from the Beekman convolution formula, results on the probability of ruin for each of these classes.

Let us start by first defining the following classes of functions:

Definition 32 (Slowly Varying Functions) L, a positive Lebesgue-measurable function on $\mathbb{R}^+$, is slowly varying at infinity ($L \in R_0$) iff

$$\forall t > 0, \quad \lim_{x \to +\infty} \frac{L(tx)}{L(x)} = 1.$$

Definition 33 (Regularly Varying Functions) L, a positive Lebesgue-measurable function on $\mathbb{R}^+$, is regularly varying at infinity with index α ($L \in R_\alpha$) iff

$$\forall t > 0, \quad \lim_{x \to +\infty} \frac{L(tx)}{L(x)} = t^\alpha.$$

A law with d.f. F is called a *law with slowly, resp. regularly, varying tails* if the survival function $1 - F$ is a slowly, resp. regularly, varying function.

The following proposition states without proof that these families are preserved under convolution.

Proposition 34 (Stability Under Convolution) *Let F_1, F_2 be two d.f.'s such that $\exists \alpha \geq 0$ such that $\forall i \in \{1, 2\}$,*

$$\begin{cases} 1 - F_i(x) = x^{-\alpha} L_i(x) \\ L_i \in R_0 \end{cases}$$

Then

$$G = F_1 * F_2$$

satisfies

$$1 - G(x) \sim_{+\infty} x^{-\alpha} \big(L_1(x) + L_2(x) \big).$$

Hence, G is regularly varying with index $-\alpha$.

A corollary follows immediately:

Corollary 35 *If $1 - F(x) = x^{-\alpha} L(x)$, for $\alpha \geq 0$ and $L \in R_0$, then for all $n \geq 1$*

$$1 - F^{*n}(x) \sim_{\infty} n \big(1 - F(x) \big).$$

This corollary serves to highlight the following interesting result.
Let

$$M_n = \max(X_i)_{1 \leq i \leq n}$$

be the biggest of n claims that have occurred. Then

$$\begin{aligned}
P(M_n > x) &= 1 - F^n(x) \\
&= \big(1 - F(x)\big) \times \sum_{k=0}^{n-1} F^k(x) \\
&\sim_{\infty} n \big(1 - F(x)\big).
\end{aligned}$$

Thus, when the individual claim size has a regularly varying tail, the aggregate amount of n claims behaves asymptotically like the maximum claim among these n occurrences. This corresponds to the idea that, when a law is fat-tailed, "big" events are sufficiently "frequent" to impose their behavior on that of the total claim amount.

This corollary, together with the Beekman convolution formula, also leads to the proof of the following result.

Proposition 36 (An Equivalent of the Probability of Ruin) *Within the framework of the Cramer-Lundberg model, if*

$$F_I(x) = \frac{1}{\mu} \int_0^x \left(1 - F(y)\right) dy \in R_{-\alpha},$$

then

$$\psi(u) \sim_{u \to \infty} \frac{1}{\theta\mu} \int_u^\infty \left(1 - F(y)\right) dy.$$

Proof The Beekman convolution formula can be written:

$$\psi(u) = \frac{\theta}{1+\theta} \sum_{n=0}^\infty (1+\theta)^{-n} \left(1 - F_I^{*n}(u)\right),$$

equivalently

$$\frac{\psi(u)}{1 - F_I(u)} = \frac{\theta}{1+\theta} \sum_{n=0}^\infty (1+\theta)^{-n} \frac{(1 - F_I^{*n}(u))}{1 - F_I(u)}.$$

Stability under convolutions and the dominated convergence theorem then imply that when $u \to \infty$, the right-hand term has a limit

$$\frac{\theta}{1+\theta} \sum_{n=0}^\infty (1+\theta)^{-n} n = \frac{1}{\theta}. \qquad \square$$

In particular, this result applies when the individual claim size variables are distributed according to the Pareto, Burr or log-Gamma law. This result also applies for a larger family of fat-tailed laws defined as follows.

Definition 37 (Sub-exponential Laws) A d.f. F with support $(0, \infty)$ is sub-exponential if

$$\forall n \geq 2, \quad \lim_{x \to \infty} \frac{1 - F^{*n}(x)}{1 - F(x)} = n.$$

We write $F \in \mathcal{S}$.

If the distribution of the claim size is lognormal, and follows a Benktander type-1 or -2, or a Weibull law with $0 < \tau < 1$, then F_I is sub-exponential.

The two preceding results apply to the sub-exponential case:

- $\forall n \geq 2, P(S_n > x) \sim_\infty P(M_n > x)$
- $\psi(u) \sim_{u \to \infty} \frac{1}{\theta\mu} \int_u^\infty (1 - F(y)) dy.$

Assuming stability under convolutions, the proofs are the same as in the case of the regularly varying tails.

As the following proposition (stated without proof) shows, this last result is also a characterization of sub-exponential laws:

Proposition 38 (Characterization of Sub-exponentiality) *Within the framework of the Cramer-Lundberg model, the following three assertions are equivalent*:

1. $F_I \in \mathcal{S}$
2. $1 - \psi \in \mathcal{S}$
3. $\lim_{u \to \infty} \frac{\psi(u)}{1 - F_I(u)} = \frac{1}{\theta}$.

We finally state three sufficient conditions for F_I to be sub-exponential.

Proposition 39 (Sufficient Condition 1 for Sub-exponentiality) $F_I \in \mathcal{S}$ *whenever F has a finite mean and*

$$\limsup \frac{1 - F(\frac{x}{2})}{1 - F(x)} < \infty.$$

Proposition 40 (Sufficient Condition 2 for Sub-exponentiality) $F_I \in \mathcal{S}$ *whenever*

$$\limsup \frac{xf(x)}{1 - F(x)} < \infty.$$

Proposition 41 (Sufficient Condition 3 for Sub-exponentiality) $F_I \in \mathcal{S}$ *whenever*

$$\limsup \frac{f(x)}{1 - F(x)} = 0,$$

and

$$\limsup \frac{xf(x)}{1 - F(x)} = \infty,$$

and

$$\limsup \frac{-xf(x)}{(1 - F(x)) \ln(1 - F(x))} < 1.$$

Chapter 2
Reinsurance Market Practices

This part describes the main characteristics of the reinsurance market. After recalling some of the institutional and legal context, we review the mechanisms of the most common reinsurance treaties.

2.1 A Legal Approach to Reinsurance

Definition of Reinsurance   Picard and Besson (1982) give the following definition of reinsurance.

> A reinsurance operation is a contractual arrangement between a reinsurer and a professional insurer (called cedant), who alone is fully responsible to the policy holder, under which, in return for remuneration, the former bears all or part of the risks assumed by the latter and agrees to reimburse according to specified conditions all or part of the sums due or paid by the latter to the insured in case of claims.

This definition warrants two remarks on the legal nature of reinsurance.

1. It first shows that, in return for remuneration, the reinsurer makes a resolute commitment to the risk borne by the cedant. Therefore, reinsurance is insurance for the insurer. More precisely, it is an insurance for insurance since it is in this fashion that the cedant covers the evolution of his commitments to the insured. Reinsurance allows him to bring his gross underwriting to the retention level imposed by his capital.
2. It then shows that insurance and reinsurance activities, close on an economic plane, are clearly distinct on a legal one.

The question of whether or not reinsurance is insurance has given rise to numerous theoretical developments. It is all the more delicate since no definition of insurance business is given by the positive laws of member states of the European Union or by European law. Under French law, on the one hand, reinsurance treaties are no insurance contracts, and on the other, reinsurers are not insurers.

G. Deelstra, G. Plantin, *Risk Theory and Reinsurance*, EAA Series,
DOI 10.1007/978-1-4471-5568-3_2, © Springer-Verlag London 2014

Reinsurance treaties are not insurance contracts. In France, article L 111-1 of the insurance Code specifies that reinsurance operations are excluded from the scope of insurance contract law. The same holds under most foreign laws.

Reinsurers are not insurers. They are not insurers of insurers. Indeed, in France, reinsurers are subject to far lighter licensing procedures and prudential rules than primary insurers and, in some countries, they are not even regulated. They are not insurers of the insured. Article L 111-3 of the insurance Code stipulates that the cedant remains alone liable to the insured.

However since 1995, French companies exclusively dedicated to reinsurance are required to comply with the accounting and financial management rules covered by the insurance Code and are subject to supervision. This convergence of the regulatory regime for insurers and reinsurers was desired by the profession in order to improve underwriting conditions abroad, in particular in the United States.

2.2 Some Data on Reinsurance

The reinsurance market is highly integrated at the global level and essentially over-the-counter. Related economic information is therefore hard to collect and relatively rare.

National data consolidation poses several problems:

– because of widely differing counting rules, figures cannot be readily compared and aggregated;
– group consolidation problems: a subsidiary cannot be accounted for both in its country of origin and in its parent company's books;
– numerous groups have international reinsurance programs, an entity first accepting risks internally, then retroceding them. It may be difficult to avoid accounting for these operations twice when in fact there is only one risk transfer.

Here, we rely on a study conducted by the Swiss Reinsurance Company on 2011 activities and published in "The essential guide to reinsurance".[1] The main features of the reinsurance market can be summarized as follows.

2.2.1 Cession Rates

In 2011, the premiums ceded by primary insurers amounted to 223 billion US$, of which 170 billion in non-life and 53 billion in life assurance.

These global cessions are broken down by geographic zones as follows in Table 2.1.

---

[1]Downloadable from www.swissre.com.

Table 2.1 Global cessions by geographic zones

	Share in global life reinsurance	Share in global non-life reinsurance
North America	59 %	38 %
Europe	24 %	30 %
Asia-Oceania	14 %	25 %
Latin America	2 %	4 %
Africa	2 %	3 %

Table 2.2 Market shares in 2011

	Market shares
Top 4	38.5 %
Top 5–10	17.5 %
Others	44 %

Remark that the highest cessions can be found in North America. This can be explained, on one hand, by the huge size of the company insurance market there, and, on the other hand, by the fact that North America is heavily exposed to natural catastrophes and liability risks.

2.2.2 Reinsurance Supply

The reinsurance industry has always been much more concentrated than that of direct insurance. This trend has accelerated since the beginning of the 90s. Today, there are about 200 companies offering reinsurance. The top ten non-life reinsurers by premium volume account for about half of the global premium volume, whereas the ten biggest life reinsurers account for about two thirds of the market. Tables 2.2 and 2.3 give the situation of market shares of reinsurers in 2011.

Historically, Germany, the US and Switzerland are the three most important domiciles for reinsurance companies. However, Lloyds of the London market is also a key player and an important supply has been developed in Bermuda during the last 25 years. Initially, monoline companies, specialized in natural disasters, were established there for tax and prudential purposes. A number of them are becoming general reinsurers with significant business activities.

More precisely, market shares of the 10 world's biggest reinsurers in 2011 can be found in Table 2.3.

Table 2.3 World's biggest reinsurers in 2011

Reinsurer	Market shares
Munich Re	15 %
Swiss Re	9.5 %
Hannover Re	7 %
Berkshire Hathaway	7 %
Lloyds	4.5 %
SCOR	4.5 %
RGA	3.5 %
Partner Re	2 %
Transatlantic Re	1.5 %
Everest Re	1.5 %

2.3 Methods of Reinsurance

The three methods of reinsurance are facultative reinsurance, facultative-obligatory (open-cover) reinsurance, as well as obligatory reinsurance. These labels characterize the rights and duties of the contracting parties.

Facultative reinsurance is historically the first form of cover. The object of the treaty is a given risk, already analyzed by the insurer who forwards his analysis to potential reinsurers. The cedant proposes a cover for this risk which can be accepted in totality or in part by one or many reinsurers of the marketplace.

Facultative-obligatory reinsurance breaks this symmetry between the freedom to cede and to accept to the detriment of the reinsurer. Its purpose is the risk cover of a category or a sub-category during a given period. Whenever the insurer is confronted to a claim falling within the scope of the treaty, he chooses to cede part of it to the reinsurer thus bound or not to do so. In this sense, for him the treaty remains optional. On the contrary the reinsurer is obliged to accept the cessions decided by his cedant. The treaty is in this sense obligatory for him.

Obligatory reinsurance restores symmetry between the contracting parties. It is the most used form of reinsurance, and so the term "obligatory" is often dropped for these treaties. During such operations, the cedant agrees to cede according to given procedures all or part of the risks falling in a given category or sub-category during a given period, very often equal to an accounting period. The reinsurer is obliged to accept all the cessions that are proposed to him under these conditions.

2.4 Kinds of Reinsurance

Though the reinsurance treaty comes under general contract law and is therefore poorly regulated, market practices confine reinsurance within a customary and fairly strict framework making a systematic classification of treaties possible. The main difference is between so-called "proportional" reinsurance and "non-proportional" reinsurance.

Before going into technical details about the various mechanisms, we list the clauses common to all reinsurance treaties, whatever be their nature.

Definition of Risks   Similarly to an insurance contract, a reinsurance treaty must first clearly define the risks whose occurrence could trigger payments from the reinsurer.

The direct insurance portfolio involved needs to be defined as follows:

– The technical nature of risks covered (e.g.: civil liability, vehicle, fire, hale, . . .)
– The geographic location of the risks covered (e.g.: the whole world, mainland France, . . .)
– The coverage period (often a calendar year, sometimes several). This point is especially crucial. For some claims, for example concerning professional civil liability, the dates of occurrence of the events giving rise to them and of the insured's claim declaration may be very far apart. It is, for example, the case for claims deriving from illnesses provoked by asbestos. If the treaty is on a *claims made* basis, the claims declared during the period of cover fall within the scope of the treaty. If the treaty functions on an *occurrence basis,* the claims that have occurred during the period of cover are covered by the treaty. Though the purpose of this book is to present the technical aspects of reinsurance, it should be stressed that the legal definition of the scope of the treaty requires special attention. This definition must be consistent with the reinsured insurance contract clauses in order to guard the ceding company against "coverage holes".

Reinsurer's Information   For a newcomer in reinsurance, the contrast between the importance of the financial amounts at stake in a reinsurance treaty and the relatively little amount of information that ceding companies provide their reinsurers, and which is rarely audited, is striking. Reinsurers pay significant amounts on the basis of sometimes brief claims reporting sheets. Observers emphasize that mutual trust between parties is an essential element in reinsurance. In fact, this "trust" is mainly sustained by the restricted feature of reinsurance supply. Few actors offer such coverage. A ceding company's inappropriate behaviour towards one of them would be quickly publicized in the market and would make it hard to renew treaties at the following maturity date. In such an event, the cedant's management would encounter problems, making future career paths difficult.

Though of relatively little importance in general, the cedant's obligations regarding information for reinsurers are precisely stated in the treaties. For new portfolios, either rapidly evolving or consisting of major risks, the composition of the portfolio must be provided to the reinsurer. This is however not the case for the "traditional" guarantees of private individuals (vehicle, damages to property). Claims, either their total costs, or only the most significant among them (whose estimation in practice exceeds a certain threshold) must be periodically declared to the reinsurers, in general monthly or quarterly.

Payment Procedures   The cedant receives premiums and pays out claim amounts daily during the accounting period. Replicating these cash flows in real time for the ceded part would generate tremendous management costs. In order to arbitrate between these management costs and the equilibrium of the treaty regarding the movement of funds, the frequency of premium payments and ceded claims is fixed in the treaty. It is also in general monthly or quarterly.

Commitment Guarantee   In several branches, the delay between filing a claim and its final settlement by the cedant is about 2 to 3 years. For the biggest claims, which are in general the most reinsured, this delay can even be much longer. During this time, the cedant has a receivable with the reinsurer, equal to the probable ceded amount taking into account the evaluation that can be made of the claim. The cedant remaining liable for the totality of the claim to the insured, treaties in general provide the procedures whereby the reinsurers guarantee their commitments so that the solvency of the cedant does not directly depend on that of the reinsurers. In France, they are of three types:

1. The reinsurer can make a "cash deposit", in practice this means providing the cedant a loan whose reimbursement is contingent on the settlement of its liabilities towards the insured.
2. He can also pledge securities (in general high rate obligations) in favour of the insured.
3. Finally, he can provide a bank guarantee.

Dispute Resolutions   Like all commercial contracts, the reinsurance treaty states the procedures for settling the disputes that may occur: the procedure for appointing arbitrators for the arbitration phase, the jurisdiction if they need to be brought before the courts. Conflicts between national laws must be taken account of very carefully during the review of these clauses since reinsurance treaties often involve companies of several nationalities.

Audit   Most treaties lay down how reinsurers can conduct audits in the cedant's premises. These clauses are rarely used in practice. It happens that reinsurers intervene directly in the management of a claim and in this case, have access to the full information in the hands of the cedant. However, most of the time, the cedant fully consents to this as it wishes to benefit from the reinsurers' expertise regarding the management of exceptional events.

Profit Sharing   Reinsurance treaties very often provide a clause giving the cedant a share in the reinsurers' profits. Under its most basic form, a certain percentage of the final result of the treaty is to be retroceded to the cedant if the balance is positive. This clause is in general multi-year and provides an unlimited carry forward of losses. An experience account aggregates the results, both positive and negative, of preceding accounting periods and the positive results first nullify past losses before the possible residual profit is returned to the cedant.

Another form of profit sharing sometimes prevalent in non-proportional reinsurance is the "no claims bonus". A fixed sum is returned to the ceding company if no event generating a payment from the reinsurer has affected the treaty.

These clauses are common to all treaties, whatever their nature. The details of the conditions under which the reinsurer intervenes determine their nature.

To describe the mechanisms of the main reinsurance treaties, we consider a direct insurance portfolio consisting of n risks. For each risk in $\{1,\ldots,n\}$, P_i is the insurance premium, S_i the annual total claim amount. We decompose S_i in the following classic manner:

$$S_i = \sum_{j=1}^{N_i} Y_{ij},$$

where N_i is the annual number of claims related to the risk i and $(Y_{ij})_{1\leq j\leq N_i}$ their claim sizes.

Finally set

$$S = \sum_{i=1}^{n} S_i \quad \text{and} \quad P = \sum_{i=1}^{n} P_i$$

to be the annual total claim amount and the total premium.

2.4.1 Proportional Reinsurance

Proportional reinsurances treaties are so named because they are constructed in order that

$$\frac{\text{Ceded Premiums}}{\text{Gross Premiums}} = \frac{\text{Ceded Claims}}{\text{Gross Claims}}.$$

The ceded premium and claim rates are equal. The two kinds of proportional treaties are the *quota-share* treaty and the *surplus* treaty.

(i) Quota-Share

This is the simplest reinsurance treaty. The reinsurer cedes a percentage $(1-a)$ of his premiums as well as of his gross claims. $(1-a)$ is the cession rate, and a the retention rate (see Table 2.4).

An advantage of this treaty is the ease with which it can be implemented and handled.

With a quota-share, the cedant and the reinsurer have exactly the same ratio $\frac{S}{P}$. This property is double-edged.

• Problems of moral hazards are alleviated: insurer and reinsurer have perfectly congruent interests and the fact of being covered should not encourage the cedant to adopt a behaviour detrimental to the reinsurer if his retention rate is sufficient.

Table 2.4 Quota-share

Risk	Premiums	Claims
Total	$P = \sum_{i=1}^{n} P_i$	$S = \sum_{i=1}^{n} S_i$
Retained	aP	aS
Ceded	$(1-a)P$	$(1-a)S$

- This similarity of outcome is not in general the most efficient way to reduce the volatility of the net portfolio. The treaties given subsequently break this symmetry by leaving in general the most "risky" part of the claims to the reinsurer.

Reinsurance Commission

As it stands, such a treaty would not be fair. Indeed, the insurer remains in charge of the commercial development and the management of the entire portfolio, including the ceded part. His acquisition and management expenses, for which he remains alone liable, are theoretically covered by pure premium loading. Management is simple and very light for the reinsurer. It is therefore unfair that he should collect the total loading on ceded premiums.

The reinsurance commission corrects this drawback. The reinsurer compensates the cedant for his management by retroceding to it a percentage c of the ceded premiums, called the commission rate. Setting g to be the rate of management expenses of the cedant, net of financial products, the net result of the insurer equals with respect to the gross premium:

$$r_{net}^{cedant} = a \times \left(1 - \frac{S + gP}{P} \right) + (1 - a)(c - g)$$

$$= a \times r_{gross} + (1 - a)(c - g).$$

Thus,

- if the commission rate is equal to the cedant's rate of expenses, the treaty is integrally proportional:

$$\frac{\text{Ceded Premiums}}{\text{Gross Premiums}} = \frac{\text{Ceded Claims}}{\text{Gross Claims}} = \frac{\text{Net Result}}{\text{Gross Result}}.$$

- if it is lower, the insurer cedes more profit than he cedes business activities,
- if it is higher, by reinsuring, the insurer increases his commercial profitability (result/net activity).

In practice, in order to encourage the insurer to improve his claim experience, it is not unusual for the contractual commission rate to be a decreasing function of the gross ratio $\frac{S}{P}$. This smooths out the reinsurer's result.

Even though quota-share treaties have lost ground to the benefit of non-proportional reinsurance in the last decades, they are still widely used to finance the launching of new branches of business activities. In this case, the agreement is

Table 2.5 The surplus treaty

Risk	Premiums	Claims
Total	$P = \sum_{i=1}^{n} P_i$	$S = \sum_{i=1}^{n} S_i$
Retained	$\sum_{i=1}^{n} a_i P_i$	$\sum_{i=1}^{n} a_i S_i$
Ceded	$\sum_{i=1}^{n} (1 - a_i) P_i$	$\sum_{i=1}^{n} (1 - a_i) S_i$

a multi-year one and in general lasts 3 to 8 years. The cession rate and the commission rate decrease sharply in time. The high initial cession rate lowers the initial regulatory capital requirements of the ceding company. The initially very high rate of commission resembles in fact to funding by the reinsurer, who, in the first few years, accepts a very negative reinsurance balance, compensated by an improvement of the claim record and a commission lower than management expenses once the portfolio reaches its "cruising speed".

When they intervene in such "funding" quota-shares, reinsurers are actually like insurance "venture capitalists" supplying the initial specialized funding of innovative projects and gradually disengaging.

(ii) The Surplus Treaty

Surplus treaties apply to categories for which the insured value is defined without any ambiguity (fire, theft, deaths). It is essentially a quota-share whose cession rate is not known when the treaty is signed but is calculated on a risk-by-risk basis once business is underwritten. The precise mechanism is as follows.

For each risk i, the insurer and the reinsurer agree on the maximum gross value that the insurer can guarantee. It is the *underwriting limit* K_i. They also fix the *retention limit* C_i, or the maximum net value, which the cedant therefore remains liable for after reinsurance.

If the actual value of the insured risk is R_i, the treaty applies to i as an individual quota-share whose cession rate $1 - a_i$ is given by:

$$1 - a_i = \frac{\min((R_i - C_i)^+, (K_i - C_i)^+)}{R_i}.$$

Once these cession rates are determined, the treaty operates as a quota-share for each policy (see Table 2.5).

In practice, the underwriting limit is given as multiples of the surplus retention (which is then unequivocally called "the limit").

Risks whose insured value exceeds the underwriting limit do not fall within the scope of the treaty.

Example Suppose that a cedant's limit is 1 million euros and that it is covered by a surplus of up to 5 limits. Depending on the amount of risk, the premium and claim cession rates are then as follows in Table 2.6.

Table 2.6 Example of cession rates

Insured value (millions of euros)	≤1	1.5	2	3	4	5
Cession rate (%)	0	33	50	67	75	80

The advantage of the surplus over the quota-share is that it allows us to model the retention's risk profile with greater precision: the higher the cedant's risks, the more it cedes.

This type of treaty is however relatively little used, except for portfolios of very reduced sizes because it entails a more significant administrative management than in the quota-share case. Indeed, cession rates, hence premiums and ceded claims are determined on a policy-by-policy basis, which is unreasonably complex unless the number of risks is very small. Non-proportional reinsurance allows us to reach this profile of cessions more efficiently and with a far lighter administrative setup.

2.4.2 Non-proportional Reinsurance

As the name suggests, non-proportional reinsurance includes all the treaties which by their construction do not satisfy the property of similarity between the rates of ceded premiums and ceded claims. Before describing the main treaties, it is useful to take a closer look at the notion of event, central in non-proportional reinsurance.

The Notion of Event

Contrary to the prevailing situation in proportional reinsurance, an essential prerequisite in non-proportional reinsurance is the rigorous definition of the events that trigger the reinsurer's payments. The notion of event is central to this definition. For example, in the case of the storm guarantee, an event is the union of claims due to winds of unusual intensity incurred by the insurer in a given geographical zone during a given period, in general seventy two hours. The event as defined in the contract is therefore not the same as the weather event: a four day tempest amounts to two events in the case of reinsurance. It obviously also differs from its definition in direct insurance, where each affected policy amounts to an event.

Concerning vehicles, an accident involving several policy holders of the company will in general be considered as a single event from the point of view of reinsurance, whereas from the point of view of direct insurance, there are in general as many events as policy holders involved.

In the general case, an event is the aggregate of a set of individual claims affecting the portfolio whose technical nature, date of occurrence and location have common characteristics defined in the treaty.

Table 2.7 Excess-of-loss treaty

Risk	Premiums	Claims
Total	$P = \sum_{i=1}^{n} P_i$	$S = \sum_{i=1}^{n} S_i$
Retained	$(1-Q)P$	$\sum_{i=1}^{n} \sum_{j=1}^{Ni} (\min(Y_{ij}.b_i) + (Y_{ij} - b_i - a_i)^+)$
Ceded	QP	$\sum_{i=1}^{n} \sum_{j=1}^{Ni} \min((Y_{ij} - b_i)^+, a_i)$

(i) Excess-of-Loss Treaty

Excess-of-loss treaties are in all points identical to direct insurance contracts with a deductible and a limited guarantee of the insurer.

For an excess-of-loss treaty $aXSb$, the reinsurer's compensation is the following function of the cost x of an event falling within the guarantee of the treaty:

$$\min(\max(x - b, 0), a).$$

Thus, the reinsurer intervenes if the cost of the event is greater than b. He then pays the cost of the event subject to a deductible b, without however paying an amount greater than a. The notation $aXSb$ therefore means that the reinsurer pays at most a on the cost part exceeding b. In reinsurance terminology,

- b is the treaty priority;
- a is the treaty guarantee;
- $a + b$ is the treaty ceiling.

An event whose cost is less than b will not give rise to any payment from the reinsurer. If the total cost of an event exceeds $a + b$, apart from the first b euros, the cedant will be liable for the entire part of the claim amount exceeding $a + b$. In a more financial language, an excess-of-loss is similar to buying a call option with a strike price b and to selling a call option with a strike price $a + b$ on the claims.

Contrary to the prevailing situation in proportional reinsurance, the treaty definition does not immediately give the price of the ceded claims. These are technically evaluated with the help of the tools developed in the first part of this book and by observing the past claim experience of the portfolio. They are determined *in fine* by confronting reinsurance supply and demand.

Let us return to the example of the generic portfolio used in this part. Supposing that each category i is covered by an excess-of-loss treaty $a_i XSb_i$ for which the notion of event coincides with that of a direct insurance claim, retentions and cessions are presented in Table 2.7, where Q is the ceded premium rate.

In practice, an excess-of-loss treaty is divided into different layers. For example, a treaty $200XS20$ can be divided into four layers:

- $10XS20$
- $20XS30$
- $100XS50$
- $70XS150$

Table 2.8 Classification according to rate on line or pay back

	Rate on line	Pay back
Working layers	>15	<6.5 years
Middle layers	4 to 15	6.5 to 25 layers
Cat layers	<4	>25 years

This division makes the placement of the treaties easier as each reinsurer can choose the degree of volatility of his exposure to the company by investing more or less in each layer, the highest obviously being the most "risky" ones since they pertain to the tails of distributions. They only involve payments when a very high priority is crossed.

Two regularly used indicators to characterize an excess-of-loss layer are the rate on line and the pay-back. They are defined as follows:

$$\text{rate on line} = \frac{\text{layer price}}{\text{layer guarantee}} = \frac{1}{\text{pay back}}.$$

Thus, the pay back is the number of years the reinsurer takes to collect the premiums needed to finance the payment of the guarantee. The excess-of-loss layers can be classified according to their rate on line or pay back (see Table 2.8).

The working layers, which are the lowest ones, are bound to come into play very often, whereas catastrophe layers are rarely affected.

An important point is that most of the excess-of-loss treaties include a maximum annual liability for reinsurers, independent from the number of events that have occurred. In practice, this annual limit is expressed as a multiple of the guarantee and is called the number of reinstatements. Thus, for a treaty $20XS10$ with 2 reinstatements, the reinsurers' maximum annual liability is limited to 60 ($= 3 \times 20$). The premium initially payed by the cedant only corresponds to a liability equal to a guarantee. As the occurrence of events depletes this guarantee, the cedant pays additional premiums, called reinstatement premiums, to reinstate its guarantee, possibly until the annual guarantee limit is totally consumed away.

Our last remark is about layer indexation. To take into account the fact that claims take time to be settled, several treaties provide an indexation of the guarantee and priority on the relevant cost index (e.g.: construction cost) so that the risk of inflation is borne by the cedant and not by the reinsurer. In this case, it is necessary to ensure that the index does not increase more quickly for the upper layers of a treaty, for otherwise "coverage holes" appear between the layers.

(ii) Annual Aggregate Loss

The annual aggregate loss is an excess-of-loss for which the event is the insurer's total annual claim amount. Thus, the cessions and retentions for an aggregate $UXST$ are presented in Table 2.9.

Naturally, the annual aggregate loss may cover only one or many branches of the company.

Table 2.9 Annual aggregate loss

Risk	Premiums	Claims
Total	$P = \sum_{i=1}^{n} P_i$	$S = \sum_{i=1}^{n} S_i$
Retained	$(1 - Q)P$	$\min(S, T) + (S - T - U)^+$
Ceded	QP	$\min((S - T)^+, U)$

Table 2.10 Annual stop-loss

	Premiums	Claims
Total risk	$P = \sum_{i=1}^{n} P_i$	$S = \sum_{i=1}^{n} S_i$
Retained risk	$(1 - Q)P$	$\min(S, TP) + (S - TP - UP)^+$
Ceded risk	QP	$\min((S - TP)^+, UP)$

(iii) Annual Stop-Loss

The stop-loss is identical to the aggregate loss, the only difference being that the guarantee and the priority are not expressed in figures but in percentages of gross premiums.

Thus, for a stop-loss treaty $U\%XLT\%$, cessions and retentions are presented in Table 2.10.

In general, reinsurance treaties are underwritten before the turnover of the accounting period is known. Therefore the stop-loss has an advantage over the aggregate loss since a guarantee and priority adapted to the volume of business can be obtained by this indexation technique.

In terms of risk sharing, an aggregate or stop-loss type coverage is optimal. They are, however, relatively rarely implemented except for some categories (hail) and small size companies with very volatile claim amounts. Indeed, they lead to an important moral hazard. The cedant no longer has a direct interest in managing its loss ratio in the most efficient manner once the priority is reached since only the reinsurer profits from it.

This is why the excess-of-loss is in practice the most widely used coverage. It leads to a risk sharing that is, from the cedant's point of view, preferable to those allowing proportional reinsurance, while preserving its incitement to manage its claims as best as it can.

In practice, reinsurance plans are a mixture of several types of treaties, typically an excess-of-loss for the main categories, a stop-loss for some of them, and also for the smallest companies a quota-share that applies, if not to the entire portfolio, at least to several classes of it. In this case, the order in which proportional and non-proportional reinsurance are applied to the gross claims is obviously not neutral.

Example Within a given category, the coverage of a company is provided by a quota-share, contributing 50 % of it, and by an excess-of-loss $10XS5$. The quota share comes into play before the excess in case 1, and after it in case 2. In case 1,

a gross claim must be greater than 30 to rise above the excess ceiling, whereas in
case 2 this happens whenever the claim reaches 15.

To illustrate the computation of the premium of an excess-of-loss treaty $aXSb$,
we return to the two examples of Sect. 1.2.5.

(1) Suppose first that X follows an exponential law. Then the pure premium of an
excess-of-loss treaty $aXSb$ is given by

$$E\left[\min\left((X-b)^+,a\right)\right] = \int_b^{b+a}(x-b)\lambda e^{-\lambda x}dx + \int_{b+a}^{\infty}a\lambda e^{-\lambda x}dx$$

$$= -e^{-\lambda(b+a)}\left(2a+\frac{1}{\lambda}\right)+\frac{1}{\lambda}e^{-\lambda b}.$$

It can therefore be clearly seen that the pure premium of an excess-of-loss
treaty $aXSb$ increases with the guarantee a of the treaty. This pure premium also
increases with the priority of the treaty if λa is such that $e^{\lambda a} \le 1+2a\lambda$, otherwise the pure premium decreases. In practice, $e^{\lambda a} \ge 1+2a\lambda$, and the premium
therefore decreases with the priority b.

(2) If $\ln X \backsim N(m,\sigma^2)$, then the pure premium of an excess-of-loss treaty $aXSb$ is
given by

$$E\left[\min\left((X-b)^+,a\right)\right]$$
$$= e^{m+\sigma^2/2}\left[\Phi\left(\frac{\ln(b+a)-m-\sigma^2}{\sigma}\right)-\Phi\left(\frac{\ln b-m-\sigma^2}{\sigma}\right)\right]$$
$$-(a+b)\Phi\left(\frac{\ln(b+a)-m}{\sigma}\right)+b\Phi\left(\frac{\ln b-m}{\sigma}\right)+a.$$

It can be easily seen that this pure premium increases with the guarantee a
of the treaty since

$$\frac{\partial}{\partial a}E\left[\min\left((X-b)^+,a\right)\right] = 1-\Phi\left(\frac{\ln(b+a)-m}{\sigma}\right)\ge 0.$$

However, this premium decreases with the priority b since

$$\frac{\partial}{\partial b}E\left[\min\left((X-b)^+,a\right)\right] = \Phi\left(\frac{\ln b-m}{\sigma}\right)-\Phi\left(\frac{\ln(a+b)-m}{\sigma}\right).$$

Indeed, this derivative is negative since a distribution function is an increasing
function.

Chapter 3
Optimal Reinsurance

The purpose of this last part is to define and then to implement some procedures in order to identify an "optimal" reinsurance strategy. It is both interesting and reassuring to see that in most of the cases considered, the treaty types which turn out to be optimal are the most common in practice, namely quota-share, excess-of-loss or stop-loss.

Thus, this part synthesizes the two preceding ones in the sense that, based, in particular, on arguments from risk theory presented in the first part, it provides theoretical justifications for reinsurance market practices described in the second part.

More precisely, two types of theoretical approaches to determine an optimal reinsurance are successively considered: the microeconomic approach with fundamental results from optimal risk sharing theory, and then the actuarial approach, using some tools presented in the first part.

3.1 Optimal Risk Sharing

It is interesting to note that the reinsurance market has had a privileged status in the economics of uncertainty. Indeed, one of the pioneering articles in this field, due to Borch (1962), is written in the language of reinsurance. This article characterizes optimal risk exchanges between risk adverse agents. Though the scope of these results goes far beyond reinsurance, Borch explains that he made this choice because the reinsurance market is, according to him, the empirical situation nearest to his abstract model:

> It is really surprising that economists have overlooked the fact that the [risk-sharing issue] can be studied, almost under laboratory conditions, in the reinsurance markets.

This section presents Borch's main results (1962) related to optimal risk sharing, see also Borch (1974, 1990). The presentation, divided into two parts, follows two classic stages in economics. We first characterize optimal risk exchange situations in the Pareto sense. We then show that by opening an efficient and complete risk market, this optimum can be reached in a decentralized manner.

G. Deelstra, G. Plantin, *Risk Theory and Reinsurance*, EAA Series,
DOI 10.1007/978-1-4471-5568-3_3, © Springer-Verlag London 2014

3.1.1 Pareto Optimal Allocations

Since their original formulation, Borch's ideas have been developed within more and more elaborate formal frameworks. The most recent form is the continuous time approach, whereby agents exchange loss processes that can be modeled by stochastic processes (see, for example, Aase 1992). Our aim is to present Borch's fundamental results in their simplest form. It is, therefore, sufficient to present a static model, based on Koehl and Rochet (1994).

The economy includes two dates $t = 0, 1$. Random events are formalized by a probability space $(\Omega, \mathcal{F}, P)$. The economy consists of N agents identified by $i \in \{1, \ldots, N\}$. Each agent i consumes at time 1. He is characterized by his strictly increasing and concave von Neumann-Morgenstern utility function u_i of class C^2 on $\mathbb{R}$ and by his income at $t = 1$, which considered at $t = 0$ is a real random variable with values $x_i(\omega)$. Set $X(\omega) = \sum_{i=1}^{N} x_i(\omega)$ to be the aggregate wealth.

We first define the feasible allocations of this economy. They are the ones obtained by redistributing the initial aggregate wealth, hence the ones that do not require to have an aggregate wealth greater than this initial wealth in a given state in the world.

Definition 42 (Feasible Allocations) An allocation $(y_i(\omega))_{1 \leq i \leq N} \in (L^2)^N$ is feasible iff

$$\forall \omega \in \Omega, \quad Y(\omega) = \sum_{i=1}^{N} y_i(\omega) \leq X(\omega) = \sum_{i=1}^{N} x_i(\omega).$$

Let $\mathfrak{A}$ denote the set of feasible allocations.

Among these feasible allocations, we define those that are Pareto optimal. These are allocations such that any other allocation strictly improving the well being of at least one agent, also strictly reduces the well being of at least one agent. Formally,

Definition 43 (Pareto Optimal Feasible Allocations) $(y_i(\omega))_{1 \leq i \leq N} \in \mathfrak{A}$ is Pareto optimal or Pareto efficient iff

$$\nexists (z_i(\omega))_{1 \leq i \leq N} \in \mathfrak{A} \quad \text{s.t.} \quad \begin{cases} \forall i \in \{1, \ldots, n\}, & Eu_i(z_i) \geq Eu_i(y_i) \\ \exists i_0 \in \{1, \ldots, n\} & \text{s.t. } Eu_{i_0}(z_{i_0}) > Eu_{i_0}(y_{i_0}). \end{cases}$$

Here we state without proof the following characterization of Pareto optima of this economy.

Proposition 44 (Characterization of Pareto Optima) $(y_i(\omega))_{1 \leq i \leq N} \in \mathfrak{A}$ *is Pareto optimal iff there exists a sequence* $(\lambda_i)_{1 \leq i \leq N}$ *of strictly positive reals s.t.*

$(y_i(\omega))_{1 \le i \le N}$ *is a solution of the following program*

$$\max_{z \in \mathfrak{A}} \sum_{i=1}^{N} \lambda_i E u_i(z_i)$$

subject to

$$\forall \omega \in \Omega, \quad Z(\omega) = \sum_{i=1}^{N} z_i(\omega) \le X(\omega) = \sum_{i=1}^{N} x_i(\omega).$$

Note that there are as many constraints as states of the world ω. Applying this last proposition, Pareto optimal allocations can be easily characterized. Indeed, letting $\mu(\omega)$ be the Lagrange multiplier for the feasibility constraint of such a program corresponding to the state of the world ω, the first order condition is written

$$\forall \omega \in \Omega, \quad y_i(\omega) = u_i'^{-1}\left(\frac{\mu(\omega)}{\lambda_i}\right).$$

The significant result is that the marginal utility ratios are independent from the state of the world:

$$\forall i, j \in \{1, \dots, n\}^2, \quad \frac{u_i'(y_i)}{u_j'(y_j)} = \frac{\lambda_j}{\lambda_i}.$$

Economists sometimes refer to these classic conditions in the economics of uncertainty as *Borch's theorem*.

Concavity assumptions ensure that these relations characterize optimal allocations, the first order conditions being sufficient. Hence,

Proposition 45 (Borch's Theorem) $(y_i(\omega))_{1 \le i \le N} \in \mathfrak{A}$ *is Pareto efficient iff there exists a sequence* $(\lambda_i)_{1 \le i \le N}$ *of strictly positive reals such that*

$$\forall i, j \in \{1, \dots, n\}^2, \quad \frac{u_i'(y_i)}{u_j'(y_j)} = \frac{\lambda_j}{\lambda_i}.$$

3.1.2 Decentralization of the Optimum by an Efficient and Complete Market

Subject to certain assumptions, by opening at time 0 a market for risk exchanges between agents, a situation of Pareto optimal equilibrium can be reached. This is the case in a static model when there exists a unique state-price deflator (in the terminology of Duffie 1992), in other words a function

$$\pi : \Omega \longrightarrow \mathbb{R}$$

$$\omega \longmapsto \pi(\omega)$$

such that the price at time 0 of any income $z(\omega)$ can be written $\int \pi z d P$. In other words, $\pi(\omega)$ is the sum that needs to be spent at $t = 0$ to obtain an income of 1 euro if the state in the world ω is realized at $t = 1$.

For the purposes of this book, we state without proof that the hypothesis of no arbitrage opportunities and the completeness of the market are necessary and sufficient conditions for the existence and uniqueness of such a state-price deflator.[1]

The absence of arbitrage opportunities means that it is not possible for an agent to earn a non-negative income[2] at time 1 without spending money at time 0. In other words, becoming rich without taking risks is not possible.

The completeness of markets means that the space generated by the agents' initial incomes is equal to the space generated by the different risk sources of the economy.

Since there is a unique state-price deflator, it is easy to check that competitive equilibrium leads to a Pareto efficient allocation.

Indeed, each agent $i \in \{1, \ldots, n\}$ maximizes his utility subject to budget constraints. He therefore solves the following program:

$$\underset{y_i(.) \in L^2}{\text{Max}} \int u_i\big(y_i(\omega)\big) d P(\omega)$$

subject to

$$\int \pi(\omega) y_i(\omega) d P(\omega) = \int \pi(\omega) x_i(\omega) d P(\omega).$$

It follows that:

$$\forall \omega \in \Omega, \quad y_i(\omega) = u_i'^{-1}\big(v_i \pi(\omega)\big)$$

where v_i is the multiplier associated to the budget constraint of the individual i.

Hence, equilibrium incomes only depend on state-prices and hence on the aggregate wealth.

$$y_i = y_i(X)$$

and Borch's relations are satisfied, guaranteeing that the competitive equilibrium is Pareto optimal:

$$\forall i, j \quad \frac{u_i'(y_i)}{u_j'(y_j)} = \frac{v_i}{v_j}.$$

Such a risk exchange market can be considered as a reinsurance market between N insurers whose "gross" portfolios, in other words not reinsured, generate random incomes $(x_i(\omega))_{1 \leq i \leq N}$. Deriving Borch's relations with respect to the aggre-

---

[1]See, for example, Duffie (1992).

[2]Equivalently, positive almost surely and non zero with strictly positive probability.

gate wealth X, it follows that for all i, j:

$$\frac{y_i'}{y_j'}(X) = \frac{T_i}{T_j},$$

where: $T_k = -\frac{u_k'}{u_k''}(y)$ is the risk tolerance of k.

This more intuitive result suggests that relative income sensitivities to aggregate wealth are proportional to the agents' relative risk tolerances.

From $\sum_i y_i' = 1$, it even follows that:

$$y_i' = \frac{T_i}{\sum_j T_j}.$$

Because of this last expression, the gap between this model and the reinsurance market can be bridged.

- Firstly, for utility functions most encountered in the literature—*Constant Absolute Risk Aversion* (*CARA*)—it indeed follows that:

$$\forall i \in \{1, \ldots, n\}, \quad y_i(X) = \frac{T_i}{\sum_j T_j} \times X + Cte.$$

The existence of a quota-share treaty market suffices therefore to implement a Pareto optimum.

- Secondly, if the agent i_0's risk tolerance is higher than that of the others, if not infinite in a limiting case (the agent is then risk-neutral), then he assumes them completely:

$$y_{i_0}(X) = X + Cte, \quad \text{and} \quad y_i(X) = Cte \quad \forall i \neq i_0.$$

In this case, the equilibrium gives a fair idea of the very secure non-proportional protections of type *stop-loss or aggregate loss*.

In these two cases, the competitive equilibrium allows to reach Pareto optimal allocations by using simple contracts whose form recalls that of proportional and non-proportional treaties.

3.1.3 Other Economic Analyses of Reinsurance

Justifying reinsurance by optimal risk sharing remains a very theoretical point of view. Other more concrete economic justifications of reinsurance are:

Taxation   If the tax code is convex, more precisely, if the marginal rate of taxation on profits increases with profits, or if the carry-back[3] is limited in amount or duration, then reducing the volatility of the outcome by reinsurance increases the expected value of the profit after taxes if the cost of reinsurance (the reinsurer's loading) is not too high.

Prudential Regulations   If the cost of regulatory capital or the return demanded by shareholders to provide the regulatory capital is higher than the cost of reinsurance, then it is preferable to cede a risk rather than retaining it in the balance sheet. The cost of capital and the cost of reinsurance can differ widely if prudential regulations are very strict or simply if capital and reinsurance markets are very fragmented.

Provision of Services to Ceding Companies   Catastrophic claims are by definition rare in direct insurance portfolios whereas they are standard fare for reinsurers. Cedants, unexperienced with extreme events, can therefore benefit from reinsurers' experience.

Cedants' Management Certification   The fact that reinsurers, who are experts in risk management, accept to make long-term commitments to share the fate of a cedant, may reassure the policy holders or their representatives (brokers, prudential authorities), and investors about the capability of a cedant to select and manage a portfolio. This certification can reduce the cost of financing direct insurance.

3.2 Actuarial Approaches

The actuarial methods of reinsurance optimization presented here are classified into two families according to their level of generality. The most general ones make it possible to set out a type of optimal treaty (proportional, non-proportional, ...) within a family of very broad contracts. In other words, they aim to optimize the reinsurance "design". Given a "design" or treaty type, the least general ones make it possible to determine optimal priorities or retentions. For the former approaches, our presentation is based on Goovaerts et al. (1990) and Kaas et al. (2008), for the latter on Bühlmann (1970).

3.2.1 Optimal Design

In this part, we consider a primary insurer of a risk X which can be the gross aggregate claims or a gross individual claim. We assume that X is a positive real random variable with distribution $F_X(.)$.

For any reinsurance treaty, set:

---

[3]The transfer of past losses on tax benefits.

- P to be the ceded premiums, or the cost of the treaty;
- $I(x)$ the ceded risk to the reinsurer.

We then define the set $\mathfrak{I}$ of feasible treaties by

$$\mathfrak{I} = \{I(.) \text{ is continuous, differentiable and } I(0) = 0, 0 \le I' \le 1\}.$$

These restrictions on feasible treaties are motivated by reasons of moral hazards. Indeed, in practice, the exact cost of a claim is known only a long time after its occurrence, in particular in the case of reinsured claims, which are often long and complex to settle. A discontinuity in the reinsurance treaty could incite the cedant to manipulate the total cost so as to benefit from possible threshold effects. Similarly, $0 \le I' \le 1$ ensures that

$$x \longmapsto I(x) \quad \text{and}$$

$$x \longmapsto x - I(x)$$

are increasing functions of x. Thus, it is in the interest of both the insurer and the reinsurer to try to minimize the gross claims, which would not necessarily be the case without this assumption.

Set further

$$\mathfrak{I}_P = \{I \in \mathfrak{I}/\text{The ceded premium price is } P\},$$

$$\mathfrak{I}_\mu = \{I \in \mathfrak{I}/E(I(X)) = \mu\},$$

as well as

$$Z = X - I(X)$$

to be the retained risk after reinsurance modeling the net reinsurance (aggregate or individual) claims.

In this context, an optimization criterion for reinsurance $c(.)$ is a functional $c(F_Z)$ of the distribution of the retained risk after reinsurance. Optimizing reinsurance corresponds to looking for the feasible treaty minimizing such a criterion.

In the following, we are interested in optimization criteria coherent with the stop-loss order, namely those that lead to a preference for treaties giving rise to a minimum retained risk with respect to the stop-loss order defined in the first part of this book. Formally,

Definition 46 (Criteria Preserving the Stop-Loss Order) A criterion preserves the stop-loss order iff

$$Z_1 \ge_2 Z_2 \quad \longrightarrow \quad c(F_{Z_1}) \le c(F_{Z_2}).$$

The following proposition leads to a number of criteria preserving the stop-loss order.

Proposition 47 *If u is an increasing convex function, $c(F_Z) = Eu(Z)$ preserves the stop-loss order.*

Proof Trivial: the stop-loss order is equal to the ordering induced by all "risk averters". ☐

This immediate result provides the following criteria:

1. Maximizing an expected utility over $\mathfrak{I}_P$ preserves the stop-loss order.
2. Minimizing the variance of net claim amounts over $\mathfrak{I}_\mu$ preserves the stop-loss order.
3. When the reinsurer sets prices according to the variance or the standard deviation principle, minimizing ceded premiums over $\mathfrak{I}_\mu$ preserves the stop-loss order.
4. Within the framework of the Cramer-Lundberg model, minimizing the probability of ruin over $\mathfrak{I}_\mu$ preserves the stop-loss order if the reinsurer also sets prices according to the expected value principle.

This last result follows from the monotony of the probability of ruin when the individual claim size distributions are ordered with respect to the stop-loss order.

The following proposition provides a simple method for the classification of net claim amounts with respect to the stop-loss order.

Proposition 48 (A "Single Crossing" Property) *If*

$$EI_1(X) \geq EI_2(X),$$

$$I_1 \leq I_2 \quad in \ [0, s],$$

$$I_1 \geq I_2 \quad in \ [s, +\infty[,$$

then

$$Z_1 \geq_2 Z_2.$$

Proof As

$$EZ_1 \leq EZ_2,$$

it suffices to show that F_{Z_1} and F_{Z_2} only cross each other once.

Since for $i = 1, 2$, Z_i is increasing,

$$F_X(x) = F_{Z_i}(Z_i(x)).$$

Hence, for $x \in [0, s]$,

$$F_{Z_2}(Z_1(x)) \geq F_{Z_2}(Z_2(x)) = F_{Z_1}(Z_1(x)).$$

For $x \in [s, +\infty[$,

$$F_{Z_2}(Z_1(x)) \leq F_{Z_1}(Z_1(x)).$$ ☐

A direct and important consequence of this result is that stop-loss treaties are optimal for any criterion preserving the stop-loss order.

Proposition 49 (Optimality of the Stop-Loss Treaty) *In $\mathfrak{I}_\mu$, the treaty defined by*:

$$I_d(x) = (x - d)^+$$

where d is a solution of

$$E(X - d)^+ = \mu$$

is preferable to all the others for any criterion preserving the stop-loss order.

Proof For any treaty I in $\mathfrak{I}_\mu$,

$$I'_d \leq I' \quad \text{in } [0, d],$$
$$I'_d \geq I' \quad \text{in } [d, +\infty[.$$

Hence, $I_d(x)$ and $I(x)$ cannot cross each other more than once.   ☐

Suppose now that the gross total claim amount X follows a compound Poisson process:

$$X = \sum_{i=1}^{N} Y_i,$$

where N is a Poisson process and (Y_i) are independent equidistributed variables.

Suppose also that feasible reinsurance treaties T are required to be of the type:

$$T(n, y_1, y_2, \ldots) = \sum_{i=1}^{n} I(y_i),$$

where $I \in \mathfrak{I}_\mu$.

Proposition 50 (Optimality of the Excess-of-Loss Treaty) *In this case, for any optimization criterion preserving the stop-loss order, the optimal treaty is*:

$$T_d = \sum_{i=1}^{n} (y_i - d)^+,$$

where d is a solution of:

$$E(X - d)^+ = \mu.$$

The proof is an obvious consequence of the compatibility between the stop-loss order and convolutions. Subject to this constraint on the forms of treaties, the excess-of-loss, in practice the most common form of non-proportional reinsurance, is optimal. This restriction amounts to preventing the cedant from covering himself against frequency deviations of its gross total claim amount. It can be justified by the moral hazard. Indeed, in most branches, a somewhat loose underwriting has a prominent impact on the claim frequency, the severity of the claims being largely independent from the insurer's decisions. Letting the cedant be liable for the risk of frequency is therefore a way to discipline it.

Let us conclude this part with a short example illustrating the very relative robustness of this type of results on the optimality of non-proportional reinsurance.

Indeed, suppose that a cedant optimizes its reinsurance by solving the following program:

$$\min P$$

$$\text{subject to}$$

$$\text{Var } Z = V.$$

The insurer is therefore trying to minimize the cost of reinsurance while limiting the volatility of net claim amounts.

- If the reinsurer sets prices according to the expected value principle, this problem is dual to the one which consists in minimizing the variance over $\mathfrak{I}_\mu$. The latter is the minimization of a criterion preserving the stop-loss order. In this case, the optimal treaty is, therefore, a stop-loss one.
- If the reinsurer sets prices according to the variance principle, the problem becomes

$$\min \text{Var } I(X)$$

$$\text{subject to}$$

$$\text{Var}\big(X - I(X)\big) = V.$$

The constraint can be rewritten as

$$\text{Var } I(X) = V + \text{Var } X - 2\,\text{cov}\big(X, X - I(X)\big).$$

The solution of the program is, therefore, the treaty maximizing the covariance of X and $X - I(X)$.

Hence, it necessarily follows that:

$$I(X) = \gamma + \beta X.$$

As $I(0) = 0$ by assumption,

$$I(X) = \left(1 - \sqrt{\frac{V}{\text{Var } X}}\right) \times X,$$

and this time the optimal treaty is a quota-share one.

The optimal design, therefore, depends closely on the pricing principle adopted by the reinsurer. However, in practice, the pricing process can rarely be unequivocally assimilated to one of the principles discussed in this book. A practical approach may be to find the principle that best corresponds to the price provided by the reinsurance market by observing market prices, and then to deduce which is the best optimal protection.

3.2.2 Optimal Priority and Retention

In this part, we assume given the reinsurance treaty design and we apply optimization criteria in order to find an optimal retention in the case of proportional reinsurance, and an optimal priority in the case of non-proportional reinsurance. This kind of problems goes back to de Finetti (1940).

(i) Optimal Retention for Proportional Reinsurance

A cedant has a portfolio of N independent risks with claims experiences $(S_i)_{1 \leq i \leq N}$. It can cede a proportion $(1 - a_i)$ of the risk i, where $(a_i)_{1 \leq i \leq N} \in [0, 1]^N$. Set P_i to be the premium related to the risk i, and

$$L_i = P_i - E S_i$$

the associated safety loading.

The cedant applies a mean-variance criterion: it minimizes the variance of its retained claim for a given expected value k of the net profit. It, therefore, solves the following program:

$$\min_{(a_i)_{1 \leq i \leq N}} \sum_{i=1}^{N} a_i^2 \operatorname{Var} S_i$$

subject to

$$\sum_{i=1}^{N} a_i L_i = k.$$

Setting v to be the Lagrange multiplier associated to the constraint, the 1st order conditions are:

$$\forall i \in \{1, \ldots, N\}, \quad a_i = \frac{v}{2} \times \frac{L_i}{\operatorname{Var} S_i}.$$

Thus,

- the more volatile a risk, the more it will be ceded;
- the more loaded the pure premium, the less the risk will be ceded as it is then profitable.

(ii) Optimal Priority of an Excess-of-Loss: A Static Example

Suppose now that each risk S_i follows a compound process:

$$S_i = \sum_{j=0}^{N_i} Y_{ij},$$

where N_i is the counting process and the Y_{ij}, equidistributed with d.f. F_i, are independent.

The cedant now fixes the priority M_i of each excess-of-loss treaty that applies to the risks of category i.

Set

$$Y_{ij}(M_i) = \min(Y_{ij}, M_i),$$

$$S_i(M_i) = \sum_{j=0}^{N_i} \min(Y_{ij}, M_i),$$

to be the retained individual and total claim amount of the category i.

The cedant determines the priority in each category by minimizing the net variance of its result subject to having a total expected net result equal to k.

The reinsurer sets the price of each treaty according to the expected value principle by applying a safety coefficient α_i for the category i.

Finally, set Π_i to be the premiums associated to the risk i. Then the cedant's program is formulated as

$$\min_{(M_i)_{1 \le i \le N}} \sum_{i=1}^{N} \text{Var}\big[S_i(M_i)\big]$$

subject to

$$k = \sum_{i=1}^{N} \big(\Pi_i - (1+\alpha_i)ES_i - \alpha_i ES_i(M_i)\big).$$

Before solving the cedant's program, the expected value and the variance of the net claim amounts need to be computed. The first two moments of $Y_{ij}(M_i)$ are:

$$EY_{ij}(M_i) = \int_0^{M_i} x\,dF_i(x) + M_i \int_{M_i}^{\infty} dF_i(x) = \int_0^{M_i} \big(1 - F_i(x)\big)dx,$$

$$E\big[(Y_{ij}(M_i))^2\big]$$
$$= \int_0^{M_i} x^2 dF_i(x) + M_i^2 \int_{M_i}^{\infty} dF_i(x) = M_i^2 - 2\int_0^{M_i} x F_i(x)dx.$$

By composition, it therefore follows that

$$ES_i(M_i) = EN_i \times EY_{ij}(M_i) = EN_i \int_0^{M_i} \left(1 - F_i(x)\right) dx,$$

$$\text{Var } S_i(M_i)$$
$$= \text{Var } N_i \times \left(EY_{ij}(M_i)\right)^2 + EN_i \times \text{Var } Y_{ij}(M_i)$$
$$= \text{Var } N_i \left(\int_0^{M_i} \left(1 - F_i(x)\right) dx\right)^2$$
$$+ EN_i \left(M_i^2 - 2\int_0^{M_i} x F_i(x) dx - \left(\int_0^{M_i} \left(1 - F_i(x)\right) dx\right)^2\right).$$

As

$$\frac{\partial}{\partial M_i} ES_i(M_i) = EN_i \left(1 - F_i(M_i)\right),$$

$$\frac{\partial}{\partial M_i} \text{Var } S_i(M_i) = 2EN_i \times \left(1 - F_i(M_i)\right) \times M_i$$

$$+ 2\left(1 - F_i(M_i)\right)(\text{Var } N_i - EN_i)$$

$$\times \left(\int_0^{M_i} \left(1 - F_i(x)\right) dx\right),$$

the 1st order conditions are:

$$\forall i \in [1, N], \quad M_i + \frac{(\text{Var } N_i - EN_i)}{EN_i} \left(\int_0^{M_i} \left(1 - F_i(x)\right) dx\right) = K\alpha_i,$$

where K is a constant.

In general, this equation does not have an explicit solution. However, when the counting process is a Poisson process,

$$EN_i = \text{Var } N_i \quad \text{and} \quad M_i = K\alpha_i.$$

The priority is proportional to the safety loading: the more expensive the reinsurance, the higher the retention. It is, however, interesting to note that this model predicts that priorities depend only on safety loadings, and not on distributions. If the reinsurer sets the same price for all the risks, then the priority must be uniform over the portfolio.

(iii) Optimal Priority for an Excess-of-Loss: A Dynamic Example

We conclude this section with an example determining the optimal priority in a dynamic context, that of the Cramer-Lundberg model studied in the first part. This example is based on a discrete time example of Pétauton (2000).

Recall that the Cramer-Lundberg model represents the net value U_t of an insurance company at time t, as follows:

$$U_t = u + ct - S_t$$

where

- u is the initial capital at $t = 0$.
- $S_t = \sum_{i=1}^{N_t} X_i$ is the process of the aggregate claim amounts it has payed at time t. It is a compound Poisson process. The intensity of the Poisson process is written λ. The individual claim size has expected value μ, and d.f. F. We also set $M_X(s) = E(e^{sX})$.
- $c = (1 + \eta)\lambda\mu$ is the instantaneous rate of premiums received by the company. η is the pure premium loading rate.

Recall that the Lundberg coefficient ρ is the solution of

$$1 + (1 + \eta)\mu\rho = M_X(\rho)$$

and that an upper bound for the probability of ruin is $e^{-\rho u}$.

This insurer has the possibility of ceding his excess-of-loss risks with a priority M. The reinsurer evaluates his premiums according to the expected value principle. He uses a safety coefficient η_R.

The insurer fixes his priority M so as to maximize the dividend paid (continuously) $q(M)$ subject to the Lundberg upper bound of his probability of ruin being kept below a given threshold. For a given capital, ρ is therefore constrained. The insurer maximizes:

$$q(M) = (1 + \eta - \gamma)\mu\lambda t - (1 + \eta_R)\lambda t \int_M^\infty (y - M) dF(y).$$

This is the premium rate that can be distributed, and hence the gross premium rate reduced by the net reinsurance premium rate and the rate $\gamma\mu\lambda t$ that has to be adopted to get a sufficient coefficient ρ.

γ is given by the constraint on ρ by using the definition of the Lundberg coefficient:

$$1 + \rho\mu(1 + \gamma) = \int_0^M e^{\rho y} dF(y) + e^{\rho M}\big(1 - F(M)\big).$$

Hence

$$q'(M) = \lambda t\big(1 - F(M)\big)\big(1 + \eta_R - e^{\rho M}\big).$$

This gives

$$M = \frac{\ln(1 + \eta_R)}{\rho}.$$

M being inversely proportional to ρ, it is proportional to the capital: the more wealthy a company, the higher its possible retention. For a given solvency target, it thus pays its shareholders more than its reinsurers.

3.2.3 Numerical Examples

We conclude this part by a numerical example, taken from Kaas et al. (2008), on the use of the adjustment coefficient to evaluate reinsurance treaties.

(i) Determination of a Stop-Loss Priority

Consider a portfolio whose total annual claim amount S follows a compound Poisson law with $\lambda = 1$ and $p(1) = 1/2 = p(2)$. Suppose that the safety coefficient θ is equal to 0.2 and hence that the annual premium $c = 1.8$. Let us calculate the adjustment coefficient:

$$\lambda + cr = \lambda M_X(r) \quad \Longrightarrow \quad \tilde{R} \approx 0.211.$$

We want to determine the adjustment coefficient after reinsurance and more particularly after an annual aggregate loss for which we want to fix the priority by comparing the expected values of gain and the adjustment coefficients. Suppose that the reinsurer uses the expected value principle to determine the premium with a safety coefficient $\xi = 0.8$.

As an example, we here do the calculations for a priority $d = 3$.

It can be checked (thanks to the example in Sect. 1.2.6) that the reinsurance premium is equal to

$$(1 + \xi)E\left[(S - d)^+\right] = 1.8\Pi(3) = 0.362.$$

Naturally, as the reinsurer makes a profit in expected value, the expected value of the insurer's gain decreases: this is the cost of reinsurance. The expected value of gain before reinsurance is $1.8 - 1.5 = 0.3$. After reinsurance it is $0.3 - \xi\Pi(3) = 0.139$.

We next determine the law of the gain G_i over a year. This consists of the direct premium from which the reinsurance premium and the retained claims have been deducted:

$$G_i = \begin{cases} 1.8 - 0.362 - S_i & \text{if } S_i = 0, 1, 2, 3; \\ 1.8 - 0.362 - 3 & \text{if } S_i > 3. \end{cases}$$

Table 3.1 Priority
determination

Priority	$\tilde{R}$(reins)	Expected value of gain after reinsurance
3	0.199	0.139
4	0.236	0.234
5	0.230	0.273
∞	0.211	0.300

An adjustment coefficient after reinsurance corresponds to this law

$$\tilde{R}(\text{reins}) \approx 0.199.$$

Hence, by this treaty, the adjustment coefficient has decreased. The expected value of the profit loss is too high compared to the ceded risk, and this treaty is, therefore, undesirable since it increases the Lundberg upper bound.

Similarly, for other priority values, the adjustment coefficients can be obtained after reinsurance $\tilde{R}$(reins) and so can the expected value of gain after reinsurance. See Table 3.1.

Hence, it is obvious that the decision $d = 3$ is not rational since $\tilde{R}$(reins) and the expected value of gain after reinsurance are at their lowest at the same time. For other values of d, the choice comes down to an arbitrage between a higher expected profit and a lower probability of ruin. As usual in reinsurance, it is therefore a matter of arbitrating between profitability and safety.

(ii) Proportional/Non-proportional Reinsurance

For a portfolio similar to the preceding one, we now compare two different kinds of proportional and non-proportional treaties, for which, given a gross claim x, the claims $h(x)$ the reinsurer is liable for are respectively:

$$h(x) = \alpha x \quad \text{with } 0 \le \alpha \le 1,$$
$$h(x) = (x - \beta)^+ \quad \text{with } 0 \le \beta.$$

We continue to suppose that $\lambda = 1$, $p(x) = 1/2$ for $x = 1$ and $x = 2$, and $c = 2$ (hence the cedant's safety coefficient θ is now worth $\frac{1}{3}$).

We now calculate the adjustment coefficient after reinsurance R(reins), which is the solution of the equation

$$\lambda + \big(c - c(\text{reins})\big)r = \lambda \int_0^\infty e^{r(x-h(x))} dP(x),$$

where c(reins) is the reinsurance premium to be paid to the reinsurer per unit of time. Set ξ to be the reinsurer's safety coefficient and fix, for example, $\xi = \frac{1}{3}$ and $\xi = 0.4$.

Table 3.2 Prop./non-prop.
reinsurance

		$\beta =$	2.0	1.4	0.9	0.6	0.3	0.15	0.0
		$\alpha =$	0.0	0.2	0.4	0.6	0.8	0.9	1.0
$\xi = \frac{1}{3}$	XL		0.326	0.443	0.612	0.918	1.84	3.67	∞
	QP		0.326	0.407	0.542	0.813	1.63	3.25	∞
$\xi = 0.4$	XL		0.326	0.426	0.541	0.677	0.425	*	*
	QP		0.326	0.390	0.482	0.602	0.382	*	*

In the case of proportional reinsurance (QP) $h(x) = \alpha x$, the reinsurance premium equals:

$$c(\text{reins}) = (1 + \xi)\lambda E\big[h(X)\big] = (1 + \xi)1.5\alpha,$$

so that the equation determining the adjustment coefficient after reinsurance $R(\text{reins})$ is

$$1 + \big(2 - (1 + \xi)1.5\alpha\big)r = \frac{e^{r(1-\alpha)}}{2} + \frac{e^{2r(1-\alpha)}}{2}.$$

For $\xi = \theta = 1/3$, we get $c(\text{reins}) = 2\alpha$ and $R(\text{reins}) = \frac{0.325}{1-\alpha}$; for $\xi = 0.4$ we get $c(\text{reins}) = 2.1\alpha$.

In the case of non-proportional reinsurance (XL) $h(x) = (x - \beta)^+$ with obviously $0 \le \beta \le 2$, the reinsurance premium equals:

$$c(\text{reins}) = (1 + \xi)\lambda E\big[h(X)\big] = \frac{1}{2}(1 + \xi)\big[(1 - \beta)^+ + (2 - \beta)^+\big],$$

so that the equation determining the adjustment coefficient after reinsurance $R(\text{reins})$ is given by

$$1 + \left(2 - \frac{1}{2}(1 + \xi)\big[(1 - \beta)^+ + (2 - \beta)^+\big]\right)r = \frac{e^{\min(\beta,1)r}}{2} + \frac{e^{\min(2,\beta)r}}{2}.$$

Table 3.2 gives the different results for different values of β compared to those of proportional reinsurance where α is fixed so that the reinsurer's payments are the same. Hence:

$$1.5\alpha = \frac{1}{2}\big[(1 - \beta)^+ + (2 - \beta)^+\big].$$

For $\xi = 1/3$ the reinsurer's and the insurer's safety coefficients are the same and the adjustment coefficient increases if more risks are ceded to the reinsurer.

If the reinsurer asks for a safety coefficient $\xi = 0.4$, then for $\alpha \ge 5/6$, the expected value of the retained risks $\lambda E[X - h(X)] = 1.5(1 - \alpha)$ is greater than the premium that the insurer can keep $c - c(\text{reins}) = 2 - 2.1\alpha$, and hence the probability of ruin $\psi(u) = 1$. One observes that the same holds in the case of the excess-of-loss treaty for all $\beta \leq 1/4$.

In Table 3.2, the adjustment coefficients of the excess-of-loss treaty are seen to be greater than those of proportional reinsurance. This is not surprising since among all the treaties with the same reinsurance pure premium, the excess-of-loss treaties have been seen to give the lowest probability of ruin.

References

Aase K (1992) Dynamic equilibrium and the structure premium in a reinsurance market. Geneva Pap Risk Insur Theory 17:93–136

Beard RE, Pentikäinen T, Pesonen E (1984) Risk theory, 3rd edn. Chapman & Hall, London

Beekman JA (1974) Two stochastic processes. Wiley, New York

Black F, Scholes M (1973) The pricing of options and corporate liabilities. J Polit Econ 81:637–654

Bodie Z, Merton R (2001) Finance. Prentice Hall, New York

Borch K (1962) Equilibrium in a reinsurance market. Econometrica 30:424–444

Borch K (1974) The mathematical theory of insurance. Heath, Boston

Borch K (1990) Economics of insurance. North-Holland, Amsterdam

Bowers NL, Gerber HU, Hickman JC, Jones DA, Nesbitt CJ (1986) Actuarial mathematics. Society of Actuaries, Schaumburg

Bühlmann H (1970) Mathematical methods in risk theory. Springer, Berlin

Daykin CD, Pentikäinen T, Pesonen M (1994) Practical risk theory for actuaries. Chapman & Hall, London

de Finetti B (1940) Il problema dei pieni. G Ist Ital Attuari 11:1–88

Dewatripont M, Tirole J (1994) The prudential regulation of banks. MIT Press, Cambridge

Duffie D (1992) Dynamic asset pricing. Princeton University Press, Princeton

Embrechts P, Klüppelberg C, Mikosch T (1997) Modelling extremal events. Springer, Berlin

Feller W (1969) An introduction to probability theory and its applications, vol 2. Wiley, New York

Gerber H (1979) An introduction to mathematical risk theory. Huebner foundation monographs, vol 8. Richard D Irwin Inc, Homewood

Goovaerts M, De Vylder F, Haezendonck J (1984) Insurance premiums. North-Holland, Amsterdam

Goovaerts M, Kaas R, van Heerwaarden A (1990) Effective actuarial methods. North-Holland, Amsterdam

Gouriéroux C (1999) Statistique de l'assurance. Economica, Paris

Grandell J (1991) Aspects of risk theory. Springer, Berlin

Heilmann W (1988) Fundamentals of risk theory. Verlag Versicherungswirtschaft, Karlsruhe

Hogg R, Klugman S (1984) Loss distributions. Wiley, New York

Hull JC (2000) Introduction to futures and options markets, 4th edn. Prentice-Hall, Englewood Cliffs

Kaas R, Goovaerts M (1993) Inleiding risicotheorie. Amsterdam

Kaas R, Goovaerts M, Dhaene J, Denuit M (2008) Modern actuarial risk theory: using R, 2nd edn. Springer, Berlin

Koehl PF, Rochet JC (1994) Equilibrium in a reinsurance market: introducing taxes. Geneva Pap Risk Insur Theory 19(2):101–117

G. Deelstra, G. Plantin, *Risk Theory and Reinsurance*, EAA Series,
DOI 10.1007/978-1-4471-5568-3, © Springer-Verlag London 2014

Lamberton D, Lapeyre B (1991) Introduction au calcul stochastique appliqué à la finance. Ellipses, Paris

Lundberg F (1903) Approximerad framställning av sannolikhetsfunktionen aterförsäkring av kollektivrisker. Almqvist & Wiksell, Uppsala

Merton RC (1990) Continuous-time finance. Blackwell Sci, Oxford

Mikosh T (2009) Non-life insurance mathematics: an introduction with the Poisson process, 2nd edn. Springer, Berlin

Panjer H, Willmot G (1992) Insurance risk models. Society of Actuaries, Schaumburg

Pétauton P (2000) Théorie de l'assurance dommages. Dunod, Paris

Picard M, Besson A (1982) Les assurances terrestres, 5th edn. LGDJ, Paris

Prabhu N (1980) Stochastic processes, queues, insurance risk and dams. Springer, New York

Rolski T, Schmidli H, Schmidt V, Teugels J (1998) Stochastic processes for insurance and finance. Wiley, New York

Rothschild M, Stiglitz J (1970) Increasing risk: I. A definition. J Econ Theory 3:66–84

Straub E (1988) Non life insurance mathematics. Springer, Berlin

Tse YK (2009) Nonlife actuarial models: theory, methods and evaluation. International series on actuarial science. Cambridge University Press, Cambridge

Printed in Great Britain
by Amazon